Improved Operational Testing and Evaluation

and Methods of Combining Test Information for the Stryker Family of Vehicles and Related Army Systems

Phase II Report

Panel on Operational Test Design and Evaluation
of the Interim Armored Vehicle

Committee on National Statistics

Division of Behavioral and Social Sciences and Education

NATIONAL RESEARCH COUNCIL
OF THE NATIONAL ACADEMIES

THE NATIONAL ACADEMIES PRESS
Washington, D.C.
www.nap.edu

THE NATIONAL ACADEMIES PRESS   500 Fifth Street, N.W.   Washington, DC 20001

NOTICE: The project that is the subject of this report was approved by the Governing Board of the National Research Council, whose members are drawn from the councils of the National Academy of Sciences, the National Academy of Engineering, and the Institute of Medicine. The members of the committee responsible for the report were chosen for their special competences and with regard for appropriate balance.

This study was supported by Contract No. DASW01-02-C-0011 between the National Academy of Sciences and the U.S. Department of Defense. Any opinions, findings, conclusions, or recommendations expressed in this publication are those of the author(s) and do not necessarily reflect the views of the organizations or agencies that provided support for the project.

International Standard Book Number 0-309-09102-0 (Book)
Library of Congress Catalog Card Number ISBN 0-309-52817-8 (PDF)

Additional copies of this report are available from National Academies Press, 500 Fifth Street, N.W., Lockbox 285, Washington, DC 20055; (800) 624-6242 or (202) 334-3313 (in the Washington metropolitan area); Internet, http://www.nap.edu

Printed in the United States of America
Copyright 2004 by the National Academy of Sciences. All rights reserved.

Suggested citation: National Research Council. (2004). *Improved Operational Testing and Evaluation and Methods of Combining Test Information for the Stryker Family of Vehicles and Related Army Systems*. Phase II Report, Panel on Operational Test Design and Evaluation of the Interim Armored Vehicle, Committee on National Statistics. Washington, DC: The National Academies Press.

THE NATIONAL ACADEMIES
Advisers to the Nation on Science, Engineering, and Medicine

The **National Academy of Sciences** is a private, nonprofit, self-perpetuating society of distinguished scholars engaged in scientific and engineering research, dedicated to the furtherance of science and technology and to their use for the general welfare. Upon the authority of the charter granted to it by the Congress in 1863, the Academy has a mandate that requires it to advise the federal government on scientific and technical matters. Dr. Bruce M. Alberts is president of the National Academy of Sciences.

The **National Academy of Engineering** was established in 1964, under the charter of the National Academy of Sciences, as a parallel organization of outstanding engineers. It is autonomous in its administration and in the selection of its members, sharing with the National Academy of Sciences the responsibility for advising the federal government. The National Academy of Engineering also sponsors engineering programs aimed at meeting national needs, encourages education and research, and recognizes the superior achievements of engineers. Dr. Wm. A. Wulf is president of the National Academy of Engineering.

The **Institute of Medicine** was established in 1970 by the National Academy of Sciences to secure the services of eminent members of appropriate professions in the examination of policy matters pertaining to the health of the public. The Institute acts under the responsibility given to the National Academy of Sciences by its congressional charter to be an adviser to the federal government and, upon its own initiative, to identify issues of medical care, research, and education. Dr. Harvey V. Fineberg is president of the Institute of Medicine.

The **National Research Council** was organized by the National Academy of Sciences in 1916 to associate the broad community of science and technology with the Academy's purposes of furthering knowledge and advising the federal government. Functioning in accordance with general policies determined by the Academy, the Council has become the principal operating agency of both the National Academy of Sciences and the National Academy of Engineering in providing services to the government, the public, and the scientific and engineering communities. The Council is administered jointly by both Academies and the Institute of Medicine. Dr. Bruce M. Alberts and Dr. Wm. A. Wulf are chair and vice chair, respectively, of the National Research Council.

www.national-academies.org

PANEL ON OPERATIONAL TEST DESIGN AND EVALUATION OF THE INTERIM ARMORED VEHICLE

STEPHEN M. POLLOCK *(Chair)*, Department of Industrial and Operations Engineering, University of Michigan

SETH BONDER, Consultant, Ann Arbor, Michigan

MARION BRYSON, North Tree Fire International, Marina, California

WILLIAM Q. MEEKER, Department of Statistics, Iowa State University

VIJAYAN NAIR, Department of Statistics, University of Michigan

JOHN E. ROLPH, Marshall School of Business, University of Southern California

FRIEDRICH-WILHELM SCHOLZ, The Boeing Company, Seattle, Washington

HAL S. STERN, Department of Statistics, University of California, Irvine

ALYSON G. WILSON, Los Alamos National Laboratory, Los Alamos, New Mexico

JAMES P. McGEE, *Study Director*
MICHAEL L. COHEN, *Staff Officer*
MICHAEL J. SIRI, *Project Assistant*

COMMITTEE ON NATIONAL STATISTICS
2003-2004

JOHN E. ROLPH *(Chair)*, Marshall School of Business, University of Southern California
JOSEPH G. ALTONJI, Department of Economics, Yale University
ROBERT BELL, AT&T Laboratories, Florham Park, New Jersey
LAWRENCE D. BROWN, Department of Statistics, University of Pennsylvania
ROBERT M. GROVES, Survey Research Center, University of Michigan
PAUL HOLLAND, Educational Testing Service, Princeton, New Jersey
JOEL HOROWITZ, Department of Economics, Northwestern University
WILLIAM KALSBEEK, Department of Biostatistics, University of North Carolina
ARLEEN LEIBOWITZ, School of Public Policy Research, University of California, Los Angeles
VIJAYAN NAIR, Department of Statistics, University of Michigan
DARYL PREGIBON, AT&T Laboratories, Florham Park, New Jersey
KENNETH PREWITT, School of International and Public Affairs, Columbia University
NORA CATE SCHAEFFER, Department of Sociology, University of Wisconsin, Madison

ANDREW A. WHITE, *Director*

Contents

PREFACE		ix
EXECUTIVE SUMMARY		1
1	INTRODUCTION TO COMBINING INFORMATION	11
2	EXAMPLES OF COMBINING INFORMATION	17
	Combining Information to Improve Test Design, 17	
	Combining Information to Improve Estimation, 22	
3	COMBINING INFORMATION IN PRACTICE	40
	Combining Information to Assess Suitability, Survivability, and Effectiveness, 41	
	Issues in Combining Information for Reliability Assessment, 42	
4	PREREQUISITES FOR COMBINING INFORMATION	53
	Need for a Broader Definition of Data, 54	
	Need for a Test Data Archive, 57	
	Representations, 61	
	Combining Information for Complex Systems, 65	
	Need for Additional Statistical Capabilities, 66	

5	TESTING CHALLENGES AND OPPORTUNITIES POSED BY THE FUTURE COMBAT SYSTEM	69
	Testing Challenges, 71	
	Testing Opportunities, 73	
	Strategy for Testing and Evaluation, 75	

REFERENCES   77

APPENDICES

A	Further Details Concerning the Bearing Cage Example	81
B	Technical Details on Combining Information in Estimation: A Treatment of Separate Failure Modes	85
C	The Rocket Development Program	90
D	Acronyms and Abbreviations	99

PHASE I REPORT: Operational Test Design and Evaluation of the Interim Armored Vehicle   103

BIOGRAPHICAL SKETCHES OF PANEL MEMBERS AND STAFF   213

Preface

The U.S. Army Test and Evaluation Command (ATEC) is responsible for the operational testing and evaluation of Army systems in development. As the Stryker/Stryker Brigade Combat Team (SBCT, formerly named the Interim Brigade Combat Team, IBCT) entered into the final stage of development, ATEC accelerated detailed preparations for its initial operational test (IOT). ATEC was faced with the challenge of developing a test design sophisticated enough to address the complex system of systems that Stryker/SBCT represents. Furthermore, since the reliability requirement of 1,000 miles between operational mission failures was unlikely to be demonstrated at typical levels of statistical inference based solely on the IOT, the possibility of using developmental test data jointly with operational test data needed to be considered. Cognizant that a previous National Research Council panel had issued a 1998 report stressing the need to examine models for combining information in order to address this limitation of operational test data, and considering in addition that report's examination of test design and measures issues, ATEC requested that the National Research Council form the Panel on Operational Test Design and Evaluation of the Interim Armored Vehicle (Stryker).

The charge to this panel was to explore three issues concerning the IOT plans for the Stryker/SBCT. First, the panel was asked to examine the measures selected to assess the performance and effectiveness of the Stryker/SBCT in comparison both to requirements and to the baseline system. Second, the panel was asked to review the test design for the Stryker/SBCT initial operational test to see whether it is consistent with best practices. Third, the panel was asked to identify the advantages and disadvantages of techniques for combining operational test data with data from other sources and types of use.

In a previous report (appended to the current report) the panel presented findings, conclusions, and recommendations pertaining to the first two issues: measures of performance and effectiveness, and test design. In the current report, the panel discusses techniques for combining information.

The panel was charged with a task atypical of National Research Council panels: providing an assessment and review of an ongoing activity, the operational testing of an important military system. The procedures for the extremely complex and highly sensitive testing are specified in the Army's system evaluation plan (SEP) for the Stryker family of vehicles. This panel has been able to build on the recommendations of the 1998 NRC report by treating the Stryker IOT as a case study of how the defense community might make more effective use of test resources in test design and in the analysis of test data.

In this report, the panel makes a strong argument for the use of information-combining techniques for use in the operational evaluation of Stryker and similar systems. As mentioned several times in later chapters, such techniques are sensitive to various assumptions, so that model validation is a crucial part of their proper application. In developing models, analysts will need to represent the implications of any problems or unusual events that arose during system development or developmental testing. Therefore, we strongly urge that those involved in the application of the techniques described collaborate closely with those who have in-depth knowledge of the development of the system in question.

This study is occurring at a dynamic time for the service test agencies. Defense systems are becoming increasingly complex, they are required to operate in more varied sets of environments and with greater suitability, and test budgets are increasingly limited. At the same time, new statistical methods are being developed in response to similar needs for the test and evaluation of industrial systems in development, and the ability to store and manipulate huge quantities of information is constantly improving. In these new and evolving circumstances, it is crucial for the Department of Defense to exploit state-of-the-art statistical methods that make full use of the available information for both test design and test evaluation.

The panel has greatly benefited from the excellent and generous cooperation of, and information obtained from, the staff of ATEC, in particular its previous commanding officer, Major General John Marcello, its current commanding officer, General Robert Armbruster, Frank J. Apicella (technical director of the Army Evaluation Center), Major David Rohall (Stryker IOT evaluator), Nancy Dunn, Ron Corson, and Bruce Grigg. In addition, we would like to thank Jack Arnold (General Motors, retired), Steve Daly and Ernest Seglie (both at the Office of the Director of Operational Test and Evaluation, DOT&E), Paul Ellner (Army Materiel Systems Analysis Activity), Don Gaver (Naval Postgraduate School), Chuck Hemeyer (General Motors), Max Morris (Iowa State University), and Nancy Spruill (Office of the Undersecretary of Defense for Acquisition, Technology, and Logistics) for providing presentations to the panel on the topic of combining information.

This report has been reviewed in draft form by individuals chosen for their diverse perspectives and technical expertise, in accordance with procedures approved by the NRC's Report Review Committee. The purpose of this independent review is to provide candid and critical comments that will assist the institution in making the published report as sound as possible and to ensure that the report meets institutional standards for objectivity, evidence, and responsiveness to the study charge. The review comments and draft manuscript remain confidential to protect the integrity of the deliberative process.

We thank the following individuals for their participation in the review of this report: Herman Chernoff, Professor Emeritus of Statistics, Harvard University; John D. Christie, Senior Fellow, Logistics Management Institute; Donald P. Gaver, Jr., Distinguished Professor of Operations Research, Naval Postgraduate School; and Dennis E. Smallwood, Professor of Social Science, U.S. Military Academy.

Although the reviewers listed above have provided many constructive comments and suggestions, they were not asked to endorse the conclusions or recommendations, nor did they see the final draft of the report before its release. The review of this report was overseen by William F. Eddy, Department of Statistics, Carnegie Mellon University. Appointed by the National Research Council, he was responsible for making certain that an independent examination of this report was carried out in accordance with institutional procedures and that all review comments were carefully considered. Responsibility for the final content of this report rests entirely with the authoring committee and the institution.

This report is the collective product of the entire panel, and each member took an active role in drafting sections of chapters, leading discussions, and reading and commenting on successive drafts. Staff at the NRC also made important contributions to our work. We express our appreciation to Andrew White, director of the Committee on National Statistics, for his valuable insight, guidance, and support, and to Michael Siri, the panel's project assistant, who was indispensable in organizing meetings, arranging travel, compiling agenda materials, coordinating with the interested community, copyediting and formatting the report, and managing the exchange of documentation among the committee members. Finally, the editing skills of Cameron Fletcher have improved this report in many ways.

Stephen M. Pollock, *Chair*
James P. McGee, *Study Director*
Michael L. Cohen, *Staff Officer*
Panel on Operational Test Design
   and Evaluation of the Interim
   Armored Vehicle (Stryker)

Executive Summary

This report provides an assessment of the U.S. Army's planned initial operational test and evaluation (IOT&E) of the Stryker family of vehicles. Stryker is the intended platform for the Stryker Brigade Combat Team (SBCT). The Army Test and Evaluation Command (ATEC), charged with conducting operational testing and evaluation of Army systems in development, has been asked to take on the unusual responsibility of designing and conducting operational testing and evaluation of both the vehicle and the SBCT concept and has requested the assistance of the National Research Council (NRC) in this effort.

The Panel on Operational Test Design and Evaluation of the Interim Armored Vehicle (Stryker), building on the recommendations of an earlier National Research Council report (National Research Council, 1998), considers the Stryker IOT&E to be a case study of how ATEC (and the other service test agencies) can more effectively conduct operational test design and evaluation consistent with state-of-the-art statistical principles and practices.

The panel has been asked to address three aspects of the operational test design and evaluation of Stryker: (1) the selection of measures of performance and effectiveness to be used to compare the SBCT equipped with the Stryker against the baseline force, a light infantry brigade; (2) whether the current operational test design for Stryker is consistent with state-of-the-art methods for experimental design; and (3) the advantages for evaluating Stryker, and more generally any complex defense system, through the use of information from the initial operational test combined with that from developmental tests, modeling and simulation, test data and field use of comparable systems, and engineering judgment and experience. The first two topics, measures and test design, were addressed in the panel's first phase report, which is appended to this report. The third item, combining information, is addressed in this report. This executive summary pertains to both reports.

MEASURES OF EFFECTIVENESS

The panel was asked to consider what measures of effectiveness (MOEs) would be useful for comparing Stryker against a baseline system and focused on issues such as: the disadvantages of rolling up disparate MOEs in a single overall number, the advantages of various force ratio measures, and the calibration and scaling of subjective evaluations made by subject-matter experts (SMEs). We have also pointed out the need to develop scenario-specific MOEs for noncombat missions and suggested some possible candidates. The panel concluded that no single measure could be devised for the value of situation awareness, and so approaches were proposed for collective measurement. Further, modeling and simulation were suggested for use in augmenting test data to help assess situation awareness.

With respect to determining measures of reliability, availability, and maintainability (RAM), the initial operational test will provide a relatively small amount of vehicle operating data, compared with the information obtained in training exercises and developmental testing, and thus may not be sufficient to address all of the reliability and maintainability concerns of ATEC. This lack of useful RAM information will be exacerbated by the fact that the initial operational test is to be performed without using add-on armor. For this reason, the panel stressed that RAM data collection should be an ongoing enterprise, with failure times, failure modes, and maintenance information tracked for the entire life of each vehicle (and its parts)—including data from developmental testing and training—and recorded in appropriate databases. System performance should be assessed both separately, by specific failure mode, and across failure modes, rather than assigning a single failure rate for a vehicle based on a simple exponential model for all failures. Failure propensity should be related to environmental and operational causes and conditions, including maintenance.

TEST PLANNING AND EXPERIMENTAL DESIGN

The initial proposed experimental design for Stryker risked confounding observed differences between Stryker and the baseline system with important sources of uncontrolled variation. In particular, the initial test design called for the Stryker/SBCT trials to be run at a different time of year from the baseline trials, which may have confounded time of year with a difference in effectiveness between the baseline force and the Stryker/SBCT forces. We therefore recommended that these events be scheduled as closely together in time as possible and interspersed if feasible. We have been pleased to learn that the final design of the IOT for Stryker has these test events scheduled very closely together.

In addition, we recommended that other potential sources of confounding, such as player learning and nighttime versus daytime operations, should be addressed with alternative designs. One alternative suggested to avoid confounding due to player learning was to use four separate groups of players, one for each of the two opposing forces (OPFORs), one for the Stryker/SBCT, and one for the baseline system. Alternating teams from test replication to test replication between the two systems under test would also be a reasonable way to address differences in learning, training, fatigue, and competence. The panel is pleased to note that the design the Army now proposes has addressed player learning through the use of separate player teams for the Stryker and baseline systems.

We pointed out the difficulty of identifying a single test design to address two distinct goals: (1) determining how various environmental or use factors affect Stryker's system performance with respect to dozens of measures of performance and (2) confirming a level of performance for Stryker against either a baseline system or a set of requirements. For example, the current test design, constructed primarily to compare Stryker/SBCT with the baseline, is balanced for a limited number of factors, allocating test samples to missions and environments similar to the proportion that would be expected in field use. The design precludes focusing test cases on environments in which Stryker is anticipated to have advantages over the baseline system, and it allocates a comparable number of test cases to environments for which Stryker is anticipated to provide little or no advantage. While the design may be effective in confirming that Stryker satisfies various criteria, it reduces the opportunity to understand the possible nature and magnitude of the benefit that Stryker provides in various crucial environments.

The panel therefore described some alternative approaches to operational test design, including a two-stage design—learning and confirming—and the use of small-scale pilot tests. The latter could be particularly useful in understanding the contribution of specific performance features, such as situation awareness for the Stryker system, for example, by running some test cases with the system's situation awareness capabilities intentionally degraded or turned off in order to determine their value in particular missions or scenarios.

In addition, the panel in its earlier report provided technical advice in areas such as statistical power calculations, identifying the appropriate test unit of analysis, issues involving use of SME ratings, aggregation of measures, and use of graphical methods in test evaluation.

With respect to the general system development process, the panel believes that, absent strategic considerations, a system should not be forwarded to operational testing until the system design is relatively mature. Forwarding an immature system to operational testing is an expensive way to discover errors that could have been detected in developmental testing, and it reduces the ability of the operational test to carry out its proper functions of assessing the capabilities and limitations of the mature system and confirming that it satisfies its requirements.

The panel suggested that, in the future, to assist in test design, ATEC should prepare a straw man test evaluation report (TER) well before the initial operational test is carried out. This TER should be based on fictitious data filled out using expert judgment, as if the initial operational test had been completed, and it should include examples of how a representative data set would be analyzed, models to be used to carry out the analysis, anticipated standard deviations, confidence intervals, hypothesis tests, and other summaries. The fictitious data would be based on the experience and intuition of the analysts and what they think the results of the initial operational test might look like, including how effective the new system is likely to be in various test situations. Of course, initial operational tests collect data in great detail and, for this purpose, some of that detail could be omitted—but not discarded; we discuss in Chapter 4 of this report the utility of archiving these and other data for future use.

SYSTEM EVALUATION BY COMBINING INFORMATION

This report focuses on techniques for combining information to enhance both operational test design and evaluation. The panel has concluded that, as currently planned, the number of test replications in the IOT for Stryker, a complex system of systems, will be inadequate to support hypothesis tests at the usual significance and power levels to guide the decision as to whether Stryker should be approved for full-rate production. This inadequacy is not specific to Stryker, as stated in the 1998 NRC report (National Research Council, 1998); rather, we suspect it to be true for the great majority of acquisition category (ACAT) I systems. Therefore, ATEC should seriously consider methods for augmenting information from operational testing in order to support better decision making, and also examine how information from earlier stages of system development and from analogous systems could be formally used to assist in operational test design.

Various sources and types of information could help augment the data currently collected in operational tests. These sources include developmental testing, training exercises, other less controlled uses of the system, and information obtained from both testing and field use of similar systems as well as systems with very similar components. While ATEC already makes use of some informal methods for combining such varied types of information, in particular the use of expert opinion for test design, this report focuses on the benefits of the use of more formal methods and suggests ways to implement these methods more broadly within ATEC. Of course, there are valid concerns about the comparability of data collected either in developmental testing or in uncontrolled use for prior versions of a system, and therefore the potential dangers of improper use of these methods is also discussed.

Formal methods for combining information include complete or partial pooling of data from two or more comparable sets of tests or other use, accommodating data from disparate sources using hierarchical or random effects models, and updating prior uncertainties about critical performance measures using Bayesian techniques. We stress that both formal and informal methods require the judicious selection and confirmation of underlying assumptions as well as a careful and open process by which various types of information, some of which involve subjective judgment, are gathered and combined.

To demonstrate their breadth and nature, this report presents specific examples of these methods and their applications, including their use in test design to reduce the number of test combinations needed to capture factor interactions; pooling techniques; use of existing knowledge about a Weibull parameter to enhance the precision of the assessment of a critical failure time performance measure; and their ability to incorporate uncertainty about the types and number of failure modes and associated failure rates.

This report also presents some requirements for utilizing these methods, especially with respect to data archiving, enhancing statistical capabilities within ATEC, and the necessity of using a formal process for eliciting expert judgments on system performance.

TOWARD THE FUTURE

Stryker is intended to be an integral part of a transformation to the Future Combat System (FCS) and the Future Brigade Combat Team (FBCT), whose test design and evaluation are likely to be substantially more complicated than those of the Stryker/SBCT. First, the FCS/FBCT is intended for use in a much broader array of operational missions and environments than the Stryker/SBCT. Second, it is a more complex family of systems than the Stryker/SBCT, and effective concepts, tactics, techniques, and procedures must be developed in advance of the operational test, paying particular attention to the use of the command, control, communications, computers, intelligence, surveillance, and reconnaissance (C4ISR). Third, the FCS/FBCT networking capability must be tested. Fourth, test designs will have to be effectively tailored to the evolutionary development process for the FCS/FBCT. Finally, its enhanced reliability requirements will have to be rigorously tested.

To address these challenges, we suggest in the current report that ATEC develop a parametric space of test environments that can be strategically sampled for testing. ATEC should also develop a test and evaluation data archive to support evolutionary acquisition and a strategy for supporting test design within an evolutionary acquisition framework.

LIMITATIONS

We wish to include four points related both to the limited nature of our charge and to our advice regarding measures and experimental design. First, we note that an alternative baseline system that could have taken advantage of the SBCT infrastructure could have been tested to help understand the value of Stryker without the SBCT system. Similarly, it does not seem necessary to require that only a system that could be transported as quickly as Stryker serve as a baseline for comparison.

Second, the current test compares the Stryker/SBCT system not only with a baseline system but also with the vehicles used in the baseline. For some purposes, isolating those comparisons could be important (for example, to determine Stryker's relative maneuverability in rural versus urban terrain and to examine the effects on its utility of its mobility in those environments).

The third point concerns the capacity of the current operational test design to provide adequate information on how to tactically employ the Stryker/SBCT system. For example, how should greater situation awareness be best utilized and how should it be balanced against greater vulnerability in various types of environments and against various threats? The answers to these questions do not rely on technical or statistical analyses but rather on the essential features of the test scenarios that we were not qualified to evaluate.

The fourth issue is whether the selected missions, types of terrain, and intensity of conflict are the correct choices for operational testing to support the decision on whether to pass Stryker to full-rate production. Other missions, types of terrain, intensities, and factors not included in the current test design might have an effect on the performance of Stryker, the baseline system, or both. These factors include, for example, temperature, precipitation, the density of buildings, building height, and characteristics of roads. Moreover, there are serious problems raised by the unavailability of add-on armor for the early stages of the operational test. The panel has been obligated to take the operational mode summary/missions profile (OMS/MP) as given, but it is not clear whether additional factors that might have an important effect on performance should be included as test factors.

For these reasons, our assessment of the Stryker/SBCT IOT as currently designed reflects only its statistical merits. The IOT may be deficient in other respects that may be substantially more important than the statistical aspects of the test. Therefore, even if the statistical shortcomings discussed in this report were to be mitigated, we cannot determine whether the resulting operational test design would provide sufficient information about whether Stryker should be promoted to full-rate production.

CONCLUSIONS AND RECOMMENDATIONS

We offer here several conclusions and recommendations that we believe particularly deserve high priority (additional conclusions and recommendations are discussed in the phase I report). We begin with a review of four sets of recommendations—on test measures, statistical design, data analysis, and assessment of the Stryker/SBCT operational test in a broad context—contained in our first report. After that are presented conclusions and recommendations on combining information, derived from our current report.

Recommendations on Test Measures

1. ATEC should not roll different MOEs up into a single overall MOE that tries to capture effectiveness or suitability.

2. To help in their calibration, ATEC should ask each subject-matter expert to review his or her own assessment of the Stryker IOT missions, for each scenario, immediately before he or she assesses the baseline missions (or vice versa).

3. ATEC should review the opportunities and possibilities for subject-matter experts to contribute to the collection of objective data, such as times to complete certain subtasks and distances at critical times.

4. ATEC should use the force exchange ratio (and the loss exchange ratio when appropriate), and not the relative loss ratio, as the primary mission-level MOE for analyses of engagement results.

5. ATEC should use fratricide frequency and civilian casualty frequency to measure the amount of fratricide and collateral damage in a mission.

6. ATEC should add scenario-specific measures of performance for security operations in a stable environment (SOSE) missions.

7. ATEC should add situation awareness as an explicit test condition.

8. RAM data collection should be an ongoing enterprise. ATEC should track failure and maintenance information on a vehicle or part/system basis for the entire life of the vehicle or part/system. To do this, ATEC should set up an appropriate database. Since this was probably not done with those Stryker vehicles already in existence, it should be implemented for future maintenance actions on all Stryker vehicles.

9. ATEC should analyze failure modes separately rather than trying to develop failure rates for the entire vehicle using simple exponential models.

Recommendations on Statistical Design

10. Absent strategic considerations, ATEC should not commence operational testing until the system design is mature.

11. ATEC should consider, for future test designs, relaxing some of its current rules of test design, by (a) not allocating sample size to scenarios according to the OMS/MP but instead using principles from optimal experimental design theory, (b) testing under more extreme conditions than typically will be faced in the field, (c) using information from developmental testing to improve operational test design, and (d) separating the operational test into at least two stages, learning and confirming.

12. When specific performance or capability problems arise in the early part of operational testing, ATEC should consider the use of small-scale pilot tests focused on the analysis of these problems. For example, ATEC should consider test conditions that involve using Stryker with situation awareness degraded or turned off to determine its value in particular missions.

Recommendations on Data Analysis

13. The IOT provides sparse vehicle operating data and thus may not be sufficient to address all of ATEC's reliability and maintainability concerns. The panel therefore recommends improved data collection regarding vehicle usage. In particular, ATEC should collect, separately for different failure modes, and maintain data for each vehicle over the vehicle's entire life, including training, testing, and field use.

Recommendations on Assessing the Stryker/SBCT Operational Test in a Broad Context

14. The estimation of system suitability, in particular the estimation of mean fatigue life, repair and replacement times, and the identification of failure modes, should not be the primary responsibility of operational testing, since operational testing cannot be expected to run long enough to accurately estimate these quantities. Therefore, developmental testing should give greater priority to measurement of system (operational) suitability and should be structured to provide its test events with greater operational realism.

Conclusions and Recommendations on How to Combine Information

15. ATEC should prepare a strategy for operational testing of the FCS/FBCT that will:
    - recognize the sequential nature of the testing that will be required as part of the evolutionary acquisition process for FCS,
    - recognize both the need to evaluate the family of systems and the potential need for diagnostic experimentation of operational concepts in multiple operational situations,
    - delineate relevant questions to be addressed by testing and evaluation,
    - identify the additional data (from subsequent tests) needed to address these questions, and
    - include modeling and simulation activities as an integral part of the testing and evaluation process.

16. The Department of Defense should provide the funds to establish a test data archive that will be a prerequisite for combining information for the testing and evaluation of future systems.

17. ATEC should consider ways to increase its statistical capabilities to support future use of techniques for combining information. As a first step, ATEC should consider providing all sources and types of information to a selected group of qualified statisticians in industry and academia as a case study to determine the potential advantages of combining information for operational evaluation.

1

Introduction to Combining Information

Combining information is a term that incorporates a wide range of methods and activities. It can include formal and informal methods, involve the use of qualitative or quantitative variables, and apply to both the design of data collection and analysis of the collected data. This section outlines the range of methods for combining information that can be useful when testing and evaluating a defense system and discusses their benefits and requirements.

The combining of information exemplifies the adage, "Necessity is the mother of invention." Much statistical activity has resulted from the necessity of drawing conclusions when information from a single source is not sufficient. Information combining entails more than simply viewing a collection of numbers in a common context. If all the data sets resulting in all the information to be combined were available in their most detailed versions, one could try to view the combined data set within an appropriately wider context. (This is one purpose of regression analysis, making predictions in different contexts comparable through use of covariates.) However, original data sets often are not available, or are available only in the form of derived summary statistics, or contain only an informal collection of qualitative observations, and so statisticians cannot always consider an information-combining problem as simply an exercise in estimation, especially when the data-generating mechanism is particularly complicated.

The informal use of prior information collected from various sources is a hallmark of scientific study design. Information from previous studies is often used to suggest suitable levels for study factors where changes in a response are expected to be the most pronounced; to determine suitable sample sizes to support significance testing, where previous studies may suggest an estimate of the variability that is needed to set a test's operating characteristics; and to help select the type of statistical method for data analysis.

Formal statistical models can be used to combine data from more than one study. This is common in industry, where, for example, manufacturers combine information about a vehicle part's lifetime from a variety of vehicle model years, as long as the difference in vehicle model year does not imply substantial systematic differences in stress experienced by that part. Along the same lines, experimentally determining the reliability of large systems composed of many subsystems by testing the entire system is often difficult or impossible, either practically or economically. However, data collected on the subsystems may be combined, using formal mathematical and statistical models and assumptions, to make reliability assessments of the full system.

Informal combination of information is typically carried out in an ad hoc manner by reviewing what has been learned previously and synthesizing this information for use in a current situation. Formal combination of information, on the other hand, generally involves the use of statistical models that require a number of assumptions. If the underlying assumptions are not found to be seriously violated, then formal combination of data usually builds a stronger inference than would be possible otherwise.

The most straightforward statistical approach to information combination is the pooling of information from two or more comparable studies. Such an approach may be relevant if, for example, two or more studies involve the failure rate for similar devices. Then the number of failures and operating hours from all studies may be combined to provide a single estimate, which is improved because it is based on an increased sample size. Not only is the estimate likely to be better, but its uncertainty will also be estimated more precisely.

It is often difficult to judge whether data collected from different studies are sufficiently comparable to allow them to be effectively combined by pooling. Statistical tests can be used to judge whether or not to pool data (though rejection of the null hypothesis of consistency with distributional assumptions at typical significance levels is not necessarily where the line should be drawn about whether or not to pool). For example, the assumption of normality supports concentration on two statistical tests, the equality of means and the equality of variances, to decide whether to pool several data sets thought to obey the normal distribution. Avoiding such a strong assumption about the distribution generating the data, one may instead perform a nonparametric test, though the omnibus nature of such tests makes them somewhat less effective against individual distributional forms.

Of course, it is possible that a problem may not lend itself to any method for combining information because it is not possible to identify ways of linking the information between studies. Between the extremes of being able to pool data and not finding any methods for linking studies lies the possibility of using statistical methods to combine, for disparate data sources, the appropriate parts of the available information. For example, in the case where tests of a common mean among many studies indicate that direct pooling is not appropriate, pooling of variance estimates may still be appropriate if a statistical model can be used that allows the individual studies to have different means but the same variance. Such a model yields tighter confidence intervals, on the average, than would be possible from use of the variances from each study individually. This gain is strongest in situations where the sample sizes for the individual data sets are very small.

Another example is the case where in the analysis of several sets of reliability data it is assumed, on the basis of appropriate diagnostic tests, that the data sets are distributed according to a Weibull failure time model with the same shape parameter as a previously analyzed set of data but with characteristic life parameters that vary, perhaps in a manner related to study covariates, between the data sets. In this case, information would be combined using a parameter derived from earlier, comparable studies. This use of prior information should be accompanied by alternative analyses using a spectrum of shape parameters to determine the sensitivity of the analysis results to the assumption of a common shape parameter. (As we will emphasize throughout this report, one should only make assumptions of comparability with good physical or historical justification.)

Hierarchical or random effects models represent another form of compromise between complete pooling of data and no combination, offering the potential for data-determined degrees of combining information. Consider the Weibull example above, where the choice is between a common shape parameter for all the data sets or different shape parameters for each data set. Under a random effects model one would assume that the shape parameters were a realized sample from a population of possible shape parameters. If that population has a small variance, then the shape parameters will be essentially equivalent, which is the first of the two extremes. If the population has a large variance, then the shape parameters may differ substantially, which is the second extreme. The shape parameter variance can be estimated from the data, allowing the data to determine the degree to which the different data sets reinforce one another. Hierarchical or random effects models can be used in a variety of ways; recently it has become popular to apply them (or very similar models) using a Bayesian approach (see, e.g., Gelman et al., 1995; Carlin and Louis, 1996) based on advances that facilitate computation of Bayesian estimates and the development of associated Bayesian infrastructure.

The main methodological and practical requirement for combining information is that explicit judgments or assumptions be supported. The (possibly informal) judgment might be that information from earlier studies is relevant to the design of an upcoming study; a formal mathematical or statistical assumption may be required to combine two data sources in a particular way. In either case, there are many caveats. Assuming that the value of a parameter, such as a standard deviation, is known based on earlier experiments or experience can be problematic, especially when the knowledge is based on data collected by anecdotal accounts that rely on memory. Although apparently minor violations of assumptions made to combine two data sources may, from a purist point of view, result in improper inference, one may sometimes choose a more pragmatic approach. For example, combining data sets that have slightly different means to estimate an assumed common location parameter has the effect of translating any differences in location between the two data sets into an inflated estimate of variability. In that case, an increase in effective sample size is gained at the expense of increasing the variance of some estimated parameters. Thus the estimate of the "common" mean may be compromised while accompanying confidence bounds are both, on the one hand, narrower due to the increased sample size, and, on the other hand, wider because of the increased standard deviation. The cost of such minor differences can be large when they are magnified by extrapolation. A trade-off study of this phenomenon may well be of general interest in the context of combining information.

The defense testing environment presents an opportunity to effectively and appropriately combine information. Experience with the Stryker/SBCT test and evaluation shows that operational testing (OT) alone often does not collect enough data to permit definitive conclusions. It is therefore necessary to also use data from developmental testing,[1] training, and field experience of the given system and of related systems. Combining operational test data with developmental test or other data is possible and potentially useful and effective, but it requires careful consideration of the relationships among the data sets.

There is no evidence in the Test and Evaluation Master Plan (TEMP) or any other documents or information made available to the panel that the Army Test and Evaluation Command (ATEC) intends to use formal techniques for combining information in the final Stryker operational test evaluation. This report argues for the greater use of combining information methods, including the use of subjective expert opinion, as an important part of the operational assessment of complex defense systems in development, including Stryker. As pointed out in NRC (1998) and repeated here, without the use of these methods operational tests will typically fail to provide sufficient statistical power to support the confirmatory role of significance testing in judging the extent to which requirements of defense systems have been satisfied, and consequently whether the systems should be promoted to full-rate production. To address the disconnect between the role of significance testing in operational evaluation and the inherent limitations of significance testing due to the necessarily limited information that can be collected in operational tests, the panel recommends greater use of combining information in both test design and operational evaluation of defense systems. As will be detailed and reinforced in the following chapters, this strong advocacy of the use of these methods calls for diligence and expertise in verifying that the underlying assumptions hold to an acceptable degree in order to prevent their misapplication. Since the defense acquisition process involves a number of organizations motivated by different and often competing incentives, we also stress the need to use assessment methods that help to ensure unbiased expert opinions.

In arguing for this fundamental change to the operational evaluation of defense systems, the panel is aware of its broader nonstatistical implications, which champions of these methods in the defense test and evaluation community will have to consider during implementation. Use of develop-

---

[1]Developmental testing is often typically carried out both by DoD (government) and by contractors. Because government developmental testing is usually expected to be more fully reported (and objectively summarized) than that done by contractors, the primary intent in this report is the use of government developmental testing for use in combining developmental and operational test data. When contractor testing is fully reported, the arguments provided here apply there as well.

mental test data, expert opinion, data from training exercises, and data on similar systems as part of operational evaluation will blur the boundaries of developmental and operational testing, and will clearly have potential impacts on the defense acquisition milestone system.

In addition, information-combining techniques are sensitive to various assumptions, so that model validation is a crucial part of their proper application (see, e.g., Gelman et al., 1995). In developing models, analysts will need to represent the implications of any problems or unusual events that arose during system development or developmental testing. Therefore, we strongly urge that those involved in the application of the techniques described collaborate closely with those who have in-depth knowledge of the development of the system in question.

Furthermore, the combining information methods recommended in this report are more susceptible to misapplication than the techniques currently used by ATEC. For that reason, there is an important requirement that all steps in the development of these models and in the estimation of their parameters be fully documented so that they can be formally reviewed. Although the ultimate costs and potential shortfalls of such organizational changes must be considered, the panel is pleased to see evidence that these organizational changes are already under consideration.

The remainder of this report is structured as follows. Chapter 2 provides simple examples of methods for combining information within the weapons systems test and evaluation context to suggest approaches, explain considerations, and identify potential advantages. Chapter 3 presents more realistic examples of how modeling for combining information can be applied to Army operational test and evaluation, considering the Stryker system at times as a specific application, and discusses implementation issues relating to combining information methods in the context of weapons system testing and evaluation. Chapter 4 identifies the resources, tools, and capabilities required to support the development of models for combining information in the context of defense test and evaluation. Chapter 5 discusses combining information for the operational test and evaluation of the Future Combat System (FCS)/Future Brigade Combat Team (FBCT).

We direct interested readers to the National Research Council report on *Combining Information: Statistical Issues and Opportunities for Research* (NRC, 1992), a valuable resource that provides additional technical details and useful references for methods of combining information. In addition, for other related research see Samaniego et al. (2001), Samaniego and Vestrup (1999), Arcones et al. (2002), and Gaver et al. (1997).

2

Examples of Combining Information

Prior information is critical in planning and designing efficient operational tests and in the evaluation of system performance when used in combination with information from tests. In this chapter, first we illustrate the importance of exploiting prior knowledge in the test design phase of the operational evaluation process in an example closely related to the Stryker operational test. We then discuss its use more generally in planning the test, selecting the experimental design, and selecting sample sizes for testing. Following this, we explore a variety of techniques in which prior information can be used in combination with test data to provide assessments of system performance.

COMBINING INFORMATION TO IMPROVE TEST DESIGN

In our example, a slightly simplified version of the current operational test plan for Stryker would compare the baseline and Stryker systems across a range of scenarios involving four factors, each at two levels: mission (raid vs. perimeter defense), intensity (medium vs. high), terrain (urban vs. rural), and company pair (A vs. B). A complete factorial design involving all four factors requires testing both the baseline and Stryker systems at $2^4 = 16$ combinations, for a total of 32 test cases. While this allows for estimation of the main effects and interactions of all orders, depending on availability of resources (number of test replications), it may be infeasible. Prior information about the nature and direction of the interactions would allow use of fractional factorial designs to reduce the number of test combinations. Box, Hunter, and Hunter (1978:375) observe that "there tends to be a redundancy in [full factorial designs]—redundancy in terms of an excess number of interactions that can be estimated and sometimes in an excess number of variables [components] that are studied. Fractional factorial designs exploit this redundancy."

In the example presented here, prior knowledge that the third-order interaction *mission* × *intensity* × *terrain* is not likely to be large and that *company pair* is not likely to have a strong interaction with any of the other factors would permit use of a fractional factorial experiment with eight runs (for each system) to test all of the relevant combinations. This would be a 2^{4-1} Resolution IV design in which the factor *company pair* is aliased[1] with the third-order interaction *mission* × *intensity* × *terrain*. As a consequence, the following sets of two-factor interactions are aliased with each other:

- *mission* × *intensity* with *terrain* × *company pair*
- *mission* × *terrain* with *intensity* × *company pair*
- *terrain* × *intensity* with *mission* × *company pair*

Since prior knowledge suggests that *company pair* is not likely to interact with any of the factors, the 8-run fractional factorial design presented in Table 2-1 can be used to safely estimate the three two-factor interactions of interest: *mission* × *intensity*, *mission* × *terrain*, and *terrain* × *intensity*. This achieves reduction of the total number of possible test combinations by half, saving costs and time during the operational testing phase.

Another way of using prior information to reduce the number of test replications is to use knowledge of where changes in the levels of test factors result in more substantial changes in the response under study (e.g., in the current context, the performance of a defense system). Through the adapted use of these factor levels, one can reduce the number of test replications because the response of interest is (approximately) maximized (assuming the information used is correct).

---

[1]The term "aliased" means that the linked effects are not individually estimable given the reduced set of test events, and so one estimates the sum of their joint effects. Given the assumption of *company pair* not interacting with the other factors, all but one of the aliased effects are assumed to equal zero, thereby permitting the estimation of the remaining effect.

TABLE 2-1 2^{4-1} Resolution IV Fractional Factorial Design

Run	Intensity	Mission	Terrain	Company Pair
1	Medium	Raid	Rural	A
2	Medium	Raid	Urban	B
3	Medium	PD	Rural	B
4	Medium	PD	Urban	A
5	High	Raid	Rural	B
6	High	Raid	Urban	A
7	High	PD	Rural	A
8	High	PD	Urban	B

NOTE: PD represents perimeter defense.

Test Planning

Operational testing and evaluation of military systems involve substantial resources and time, and the decisions to be made have important consequences for national security. Given the high stakes, it is critical that operational testing be planned and executed carefully and systematically and that as much relevant prior information as possible be taken into account in designing efficient test plans. It is difficult, and in some cases impossible, to generate useful information from a poorly designed test plan.

Effective test design relies on the crucial prior step of test planning. Within the statistical community there has been much more attention paid to the development of efficient techniques for the design of experiments than on the planning process that precedes it. Hahn (1993) notes:

> Experimental design is both an art and a science. The science deals with the mathematics and formalities of developing experimental plans. This is what most of the literature, including numerous articles in this journal, is about. The art of experimental design provides the framework for an effective test program that is maximally responsive . . . to the questions that the investigators wish to answer. It deals with important but seemingly non-statistical topics such as defining the goals of the [test] program, establishing the proper response and control variables, assuring proper scope and breadth, understanding the various sources of experimental error, appreciating what can and cannot be randomized, and so forth.

Related studies, subject-matter expertise, modeling and simulation, results of developmental testing, and pilot studies all play a major role in this planning process.

Many industrial organizations have recently instituted systematic processes for planning and executing large-scale experiments based on quality management principles such as six sigma. A key component of this process is the use of templates for systematic elicitation and incorporation of prior information. The process involves, for example, developing consensus in identifying key response variables, target values, and ideal functions (i.e., functions that specify the relationship between signals and responses); and documenting subject-matter knowledge and relevant background from past studies. Factors that affect the response variables are similarly identified and classified into control factors and noise variables. Subject-matter expertise or past studies are used to determine the range of values and their predicted impact on the response variables, identify constraints such as costs and the feasibility of varying the factors during experiments, and develop strategies for measuring noise variables or for introducing and systematically varying them in the experiment. Some industrial organizations make use of predesign master guide sheets (see, e.g., Coleman and Montgomery, 1993) that query the test designers to specify the objectives of the test, any relevant background issues, response variables, control variables, factors to be held constant, nuisance factors, strong interactions, any further restrictions on test inputs, design preferences, analysis and presentation techniques, and responsibility for coordination.

The systematic processes and the use of prior knowledge are also needed in selecting the design factors to be studied, their levels, and possible interactions. All of these decisions need to be made before selecting an appropriate experimental design.

Selecting the Experimental Design

There are many approaches to designing experiments. For the applications considered in this report, by far the most useful of these are factorial and fractional factorial designs (for details, see Box and Hunter, 1961). This class of experimental designs has very good statistical properties, including balance and robustness, in a broad range of situations. Full factorial designs, however, involve testing all possible combinations, which can lead to an excessive number of test scenarios when the number of factors, or levels per factor, is large. For that reason, fractional factorial designs that examine a carefully selected subset of all possible combinations of design factors are much more cost efficient. There is an extensive literature on this topic (Box, Hunter, and Hunter, 1978; Wu and Hamada, 2000). However, as mentioned above, prior information about which higher-order interactions are sufficiently small must be used when selecting appropriate fractions of the full factorial designs. Sequential follow-up strategies can verify the validity of these assumptions, although they may not be as useful in the operational test context, given the various constraints on use of military personnel, test ranges, and other resources.

There is also a large literature on so-called optimal designs. In this approach, the assumption is that the response model is known up to some parameters, and the goal is to estimate either the unknown parameters or the response surface at some design point. An illustrative example is the linear model with explanatory variables X_1 and X_2:

$$Y = \beta_0 + \beta_1 X_1 + \beta_2 X_2 + \varepsilon$$

The goal of optimal design in this example is to collect various observations of Y at specific design points (X_1, X_2) that are chosen optimally to maximize either the precision in estimating the regression coefficients (the β's) or the expected response at selected values of X_1 and X_2, assuming that the linear model is correctly specified. Other optimal designs that correspond to the maximization or minimization of other criteria of interest require prior information about the form of the model, such as the above linear model with no interaction term. In the case of a linear model, the optimal design for estimating the regression coefficients requires testing only at the extremes of the design space. While this leads to good precision if the linear model is a close approximation to the truth, the design is very nonrobust to violations of this assumption. This property of nonrobustness, more generally, is why optimal designs are not used extensively, except in cases where one is very confident about prior information. Related discussions of Bayesian optimal designs examine formal incorporation of prior information about model parameters (Chaloner, 1984).

Selecting Sample Sizes

Selection of sample sizes is dependent on the objective of the operational test. Is the objective to estimate system performance for specific types of environments of use, or to estimate the average performance across environments of use? Larger samples are needed for the former test objective. If a confirmatory hypothesis test is to be used as a basis for a decision on system promotion, the statistical power of the test against important alternative hypotheses concerning system performance (such as modestly failing to meet a requirement) needs to be computed and related to the costs and benefits of making incorrect decisions regarding promotion. The statistical power will be a function of the significance level of the hypothesis test in question, but, more importantly, it will be a function of the variance of the test statistic (e.g., average failure rate). The variance of the test statistic is not directly measured prior to carrying out the operational test; however, it can often be indirectly estimated through use of development test information, pilot studies, or variances estimated for similar systems and adjusted through the use of engineering judgment. Such indirect estimates are valuable in judging, prior to an operational test, whether the test size will be adequate to support significance testing used for this confirmatory purpose. When such an analysis suggests that test sizes sufficient for this purpose are not likely to be feasible given costs, models for combining information should be examined as a method for reducing variances.

COMBINING INFORMATION TO IMPROVE ESTIMATION

Combining Information by Pooling

It is difficult to draw useful conclusions from data sets with small sample sizes because the signal contained in the data (e.g., the difference in performance between two defense systems) is fixed, while the variability of the signal estimate is relatively high for small data sets (but decreases as the sample size increases). To address this difficulty, much ingenuity has been applied to developing methods for borrowing strength from several small samples by combining or pooling them. The methods include pooling K samples (where K is some number larger than one), pooling K samples with different means and common variances, pooling using linear or quadratic regression, and various generalizations of pooling with regression, including various nonparametric fitting algorithms and hierarchical and random effects models.

Before discussing some of these methods, we first point out that even viewing a collection of numbers as a simple random sample represents a form of combining information. The random sample model, viewing a collection of data as coming from a common distribution, is so commonly applied that it is usually not considered as relying on any assumptions, but this is not the case. The consideration of a data sample—say a group of times to first failure—as generated from a common distribution represents a form of combining information, in that individual data values are grouped into one collective, and this combining requires justification, which could include consideration of whether the data were obtained through the use of sufficiently similar processes. In addition, it would be necessary to argue that the individual data values were independently generated (or at least exchangeable). Through the empirical distribution function, such a sample provides a much better description of the underlying distribution and associated features—including the mean of the underlying distribution—than any one of the numbers by itself would be able to provide.

"Pooling samples" is most often understood to mean that one has two or more samples (in this discussion referred to as having K samples), typically of small sizes, and there are reasons to believe that these samples come from populations having the same distribution function. For example, one might have collected times to first failure for several systems in developmental testing and for an additional, smaller number of systems in operational testing. If all samples are pooled into one large sample regardless of where they came from, the required assumption is that the origin of each sample has no impact on the distribution of sample values. Diagnostic checks should be run to show that the samples do not contradict this underlying assumption. Unfortunately, when diagnostic checks are based on small samples, they tend to be somewhat forgiving; i.e., even moderate differences in the sampled populations are not easily discernible. From a pragmatic point of view, these moderate differences in the generating distributions often do not matter, but this inability to discriminate needs to be analyzed and if necessary addressed through the use of nonparametric techniques.

Diagnostic checks can include many possibilities, ranging from informal graphical box plots or pairwise quantile-quantile plots to formal parametric or nonparametric hypothesis tests. In an example of the parametric approach, we assume that the individual samples come from normal populations, and so the decision to pool depends only on whether the sample means and variances are homogeneous. This could be tested using the classical F-test for homogeneity of means and Bartlett's test for the homogeneity of variances. The assumption of normality, in addition to the assumption of the homogeneity of the first two moments, requires a check of the normality of the individual samples. In small samples such a check would reveal only gross violations.

Nonparametric tests for the homogeneity of multiple samples avoid the assumptions of normality or of other specific distributions. Examples of such tests include the Kolmogorov-Smirnov, Cramer-von Mises, and

Anderson-Darling tests as generalized to multiple samples by computing appropriate discrepancy measures that compare the empirical distribution functions of the individual samples with that of the pooled sample (see Kiefer, 1959, and Scholz and Stephens, 1987, for details). Such tests are rank tests and are sensitive to a wide range of differences in the individual empirical distribution functions, in contrast to the analysis of variance F-test for equality of means (assuming common variances and normality) and the Kruskal-Wallis rank test, which are sensitive to differences in means but can be quite weak otherwise.

In pooling K normal samples (often transformations can be used to produce data that more closely approximate a normal distribution) that have shown strong evidence of having different means, the Bartlett test can be used to check whether the samples share a common variance. If there is good evidence of homogeneous variances, one can pool them to obtain a much more accurate assessment of the common variance. This in turn has beneficial consequences for confidence intervals for the means, which, if based on the pooled variance estimate, would be narrower on the average. The benefit can be substantial when the sizes of the individual samples are small.

Sometimes the means of underlying samples vary according to functions of covariates that were observed in conjunction with each sample value. For example, the failure rate of a system might be a simple function of some measure of stress to which the systems have been exposed. Absent a model linking the various samples, one could view the sample values with common covariate values as a collection of single samples and proceed accordingly. Of course, the sample sizes at individual covariate values are likely to be extremely small. However, when a useful model can be identified, a stronger form of pooling, using multiple regression, can be exploited if one can closely approximate the means of the response of interest as a linear function of the known covariates. (The assessment of the validity of regression models has been well studied; see, e.g., Belsley et al., 1980). In particular, the residuals are useful to examine to assess conformity with assumptions of linearity, homogeneous variances, and existence of outliers. Such a model would be determined by a small number of parameters, which can be estimated using all sample values simultaneously (by the method of least squares, for example). The influence of all sample values is therefore pooled, i.e., used jointly, in estimating these few parameters. The accuracy of such estimates of the conditional means provided by the fitted values from the regression model is much greater than that afforded by just using the mean of all sample values for data collected at the covariates of interest, if they were even available. The pooling here therefore has the additional benefit of providing estimates for covariates for which no sample values were available.

In addition to this pooled (structural) model for estimating the mean function, there is the option of assuming constant variances of the sample values across all covariates. This extension of the pooling idea estimates a pooled variance from all the residuals and thus increases the degrees of freedom in the pooled variance estimate, in turn improving the accuracy assessment of the mean estimates as it is reflected in the confidence intervals. This pooling, as usual, depends on the validity of the various assumptions, and diagnostic checks including residual analyses need to be made before building on them.

Pooling using regression is a special case of a more general approach, including generalized linear models and various nonparametric fitting techniques, which can be applied to normal, count, and other forms of data. Although many textbooks on regression do not emphasize the interpretation of regression as pooling, the pooling perspective provides a strong underlying theme in discussions of regression. The pooling occurs through the use of structural models that are characterized by a few unknown parameters and that allow analysis, using covariates, of pooled data collected under various conditions. All the data simultaneously influence the model fit, and as a result more accurate estimates of the conditional means can be obtained.

Bayesian Inference with Binary Data

Dichotomous measures are relatively typical in defense testing. Success or failure of an offensive system is, for example, generally measured using assessments of the number of hits in a given number of trials. (We do not address here the point that the measure of distance from a target often may have advantages over the dichotomous measure.) Use of a Bayesian approach for dichotomous measures can be illustrated as follows: An operational test of a defense system includes 20 trials with dichotomous (success/failure) outcomes with interest in estimating the probability of failure, p. The probability of failure has been presumed to be small so that the number of failures in 20 trials is not likely to be large. For example, if the number of failures were $k = 2$, the maximum likelihood estimate of p would be 0.10, but the associated standard error would be around 0.07, leading to a very weak inferential conclusion. The option of running more test trials is assumed to be impossible due to logistical or budgetary constraints (e.g., the system is being tested under a number of scenarios, and therefore the number of replications for a given scenario is limited; or the system is sufficiently costly that testing until there were a large number of failures would be wasteful). In such a situation it might be useful and appropriate to include other information in the analysis of operational test results.

The previous discussion of pooling identifies several ways in which other information might be incorporated, if there are previous trials of a sufficiently similar system or if a statistical model (perhaps regression) could be used to render trials of other systems comparable. The current example assumes that pooling is not possible and instead considers the possibility of combining expert opinion with the results of the field trial. The example also assumes that a check with system experts suggests a consensus assessment that p is approximately 0.05 with reasonable confidence that p is no higher than 0.25 (see below for a discussion of methods that should be used to obtain such assessments).

A statistical approach for combining prior information with the test results is possible if the prior information is expressed in the form of a prior probability distribution for the unknown p. In the current example, the expert opinion (mean of .05, high percentile of .25) is consistent with a Beta(2,38) distribution with a mean .05 and almost all of its probability concentrated between 0 and .25. The prior distribution is presented as the continuous curve in Figure 2-1.

Given this prior distribution and a statistical model for the data, Bayes' Theorem produces the posterior distribution that represents the subjective probabilities for different values of p based on both the observed data and the prior information.[2] In this case it is natural to assume for the statistical model that the observed number of failures y is distributed as a binomial random variable with 20 trials, each having failure probability p. The resulting posterior distribution can provide an estimate of p and a probabilistic upper bound, or any other summary of uncertainty about p, based on the data and prior information.

---

[2]The posterior distribution is subjective even though it can be represented as a mixture of empirical frequencies.

EXAMPLES OF COMBINING INFORMATION

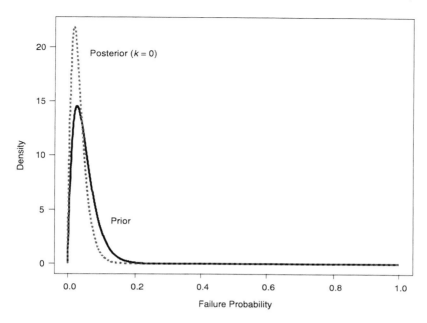

FIGURE 2-1   Prior distribution and posterior distribution (given $k = 0$).

To illustrate the approach, Table 2-2 presents, for several possible outcomes for the operational test, the conclusions one might draw by combining information. One posterior distribution is shown as the dotted line in Figure 2-1, corresponding to the case where one observes $k = 0$ failures in 20 trials. The table gives a point estimate for the median and the 95th percentile of the posterior distribution. For purposes of comparison, the table also shows the uncombined maximum likelihood point estimate for p and upper confidence limits based on the binomial model and operational test data alone.

The results illustrate the benefits of combining information. Particularly if the number of failures is small, as expected, then combining information yields sharper conclusions regarding the upper limit for the failure probability p, especially the 95 percent upper limit. In the special case where no failures are observed, the Bayesian approach yields a much more sensible point estimate as well, because an estimate of $p = 0$ is not reasonable in this

TABLE 2-2 Estimates from Combining Information and from Direct Estimation

Failures in 20 trials	Results from combining information			Results without combining information		
	Estimated failure probability (posterior mode)	50 percent point of posterior distribution	95 percent point of posterior distribution	Estimated failure probability (sample proportion)	Upper 50 percent confidence limit (binomial model)	Upper 95 percent confidence limit (binomial model)
0	.02	.03	.08	.00	.03	.14
1	.03	.05	.10	.05	.08	.22
2	.05	.06	.13	.10	.13	.28
10	.19	.20	.29	.50	.53	.70

context.³ If the observed data are not consistent with the prior information, then the conclusions regarding p will be intermediate between the two information sources.

In the current example, when there are 10 failures in 20 trials, the results from combining information suggest much lower values of p than the observed data. These results reflect the relatively strong influence of expert opinion (the experts were nearly certain that the failure probability was below .25) and emphasize both the importance of considering the sensitivity of conclusions to a range of plausible interpretations of the prior information and the danger of using prior information that is not well founded. In this situation, the prior information seems to have been inappropriate, and the process by which it was generated should be examined.

This short example demonstrates a way to quantify and combine expert opinion with observed data in a relatively simple setting. Evaluations of complex systems would require combination of data from a number of subsystems using a similar approach, as discussed below.

Combining Information for Assessing Reliability: Sensitivity Analysis Versus Probabilistic Treatment of Uncertainty in Estimating the Reliability of a Bearing Cage

In this section, we discuss different methods for combining information in estimating the reliability of a bearing cage. Abernethy et al. (1983) present field data on a bearing cage, a component in a jet engine. A population of 1,703 units had been introduced into service over time, and there had been 6 failures. The reliability goal for the bearing cage was fewer than 10 percent failing in 8,000 hours of service (in engineering notation, that means B10 life—the time at which 10 percent fail—is greater than 8,000 hours). For display purposes, units surviving for various lengths of time were grouped into intervals of 100 hours' length. Figure 2-2 is an event plot showing the structure of the available multiply-censored data, in which failures are indicated by a row ending in an asterisk (*). Figure 2-2 shows, in row 1, that 288 units were in service for about 100 hours and none

---

³In many applications, the upper confidence bound on failure probability is more important, and in this situation it would be relatively well estimated without the use of prior information.

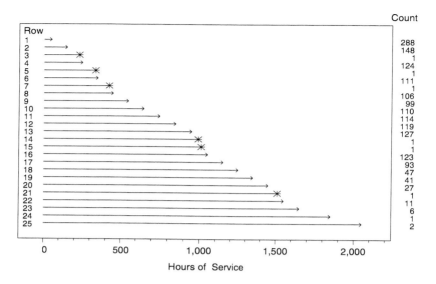

FIGURE 2-2 Event plot showing the multiply-censored bearing cage failure data. SOURCE: Abernethy et al. (1983).

experienced a failure. Proceeding to row 2, there were 148 units in service for about 200 hours and none experienced a failure. In row 3, there was a failure at around 300 hours, indicated by the asterisk. In row 4, there were 125 units in service for around 300 hours.

Figure 2-3 presents a Weibull probability plot of the same bearing cage data, showing the maximum likelihood estimate of fraction failing, the reliability target, and approximate confidence limits. The plotted points are based on nonparametric estimates (i.e., estimates computed without making any assumption about the underlying failure-time distribution) of the failure rate at each point in time. The points fall along a roughly straight line, indicating that the Weibull distribution provides a reasonable description for the failure process. The straight line through the points is the Weibull maximum likelihood estimate of the fraction failing as a function of hours in service, assuming the Weibull model is correct. The pointwise approximate 95 percent confidence limits indicate the large amount of statistical uncertainty in the estimate, owing to the small amount of information from the few failures that were observed and the extrapolation in time.

EXAMPLES OF COMBINING INFORMATION

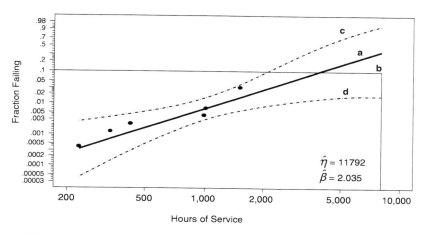

FIGURE 2-3 Weibull probability plot of the bearing cage failure data showing the Weibull maximum likelihood estimate of the fraction failing, the reliability target, and approximate pointwise 95 percent confidence limits. In the figure, the dots represent the bearing cage observed data, straight line (a) represents the maximum likelihood estimate of fraction failing, intersection (b) represents the reliability target, and curved lines (c) and (d) represent the 95 percent upper and lower pointwise confidence limits.

The point where the horizontal and vertical lines meet is the reliability target.

Since the maximum likelihood estimate of B10 life is 3,900 hours, and an approximate 95 percent confidence interval for B10 is between 2,100 and 22,100 hours, there was a concern that the B10 design life specification of 8,000 hours was not being met. On the other hand, because of the limited information in the data, it might be argued from the upper bound of the confidence interval that B10 could be as large as 22,100 hours.

Figure 2-4 is a contour plot of the Weibull relative likelihood function (a function that is proportional to the probability of the data) as a function of B10 and the Weibull shape parameter β. The maximum likelihood estimator is shown at the intersection of the horizontal and vertical lines. The probability of the data at the maximum likelihood estimate is, for example, 5 times higher than at points on the .2 contour. This function shows clearly why the upper endpoint of the B10 confidence bound is so large: small uncertainties in β are associated with a wide variety of values of B10.

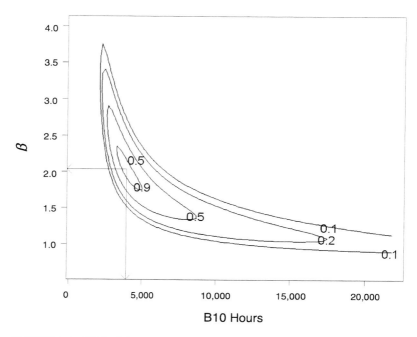

FIGURE 2-4 Weibull distribution relative likelihood for the bearing cage failure data.
SOURCE: Abernethy et al. (1983).

Abernethy et al. (1983) show that using historical or other information to fix the value of the Weibull shape parameter β reduces by a large factor the amount of statistical uncertainty in estimating B-life (quantiles) outside the range of the data. Nelson (1985) also suggests using given values for the Weibull shape parameter β when there are few failures in censored life data, but strongly encourages using sensitivity analysis to assess the effect of the uncertainty in the Weibull shape parameter because the value is never in practice known with certainty. The range of evaluation can be determined from past experience with the same failure mode in similar materials or components. A fatigue failure mechanism, because of its wearout-type behavior, would have a shape parameter greater than 1, and previous experience might suggest, for example, that β should be in the range of 1.5 to 3. Appendix A contains probability plots that are similar to Figure 2-3, but with the Weibull shape parameter β fixed at 1.5, 2, and 3.

EXAMPLES OF COMBINING INFORMATION

The overall conclusion suggested by these figures is that the bearing cage is, most likely, not meeting its reliability goal.

An alternative to the sensitivity analysis procedure is to use a prior distribution to describe engineering knowledge of the Weibull parameters. (For details, see chapter 14 of Meeker and Escobar, 1998, who use the simple graphical and simulation-based approach for Bayesian analysis suggested in Smith and Gelfand, 1992). This alternative can be illustrated by the following situation. In this example the engineers responsible for the reliability of the bearing cage have useful prior information on the Weibull shape parameter, which they quantify with a lognormal distribution with lower and upper 99 percent limits (1.5, 3). For the B10 parameter itself there is little prior information, so a diffuse prior distribution is used by specifying a loguniform distribution with lower and upper limits (500, 20,000). The Bayes rule computation of the posterior distribution involves multiplying the sampling function and the prior, and the computation can be considered a linear combination of the contours of the prior and the sample points. All inferences are based on samples generated from the posterior distribution, such as the posterior median. Figure 2-5 is a plot of the marginal posterior distribution of B10.

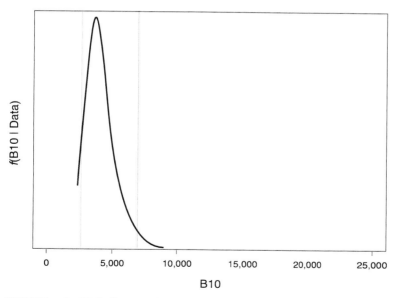

FIGURE 2-5 Weibull marginal posterior distribution for B10 of bearing cage life and 95 percent credibility intervals.

Contrasting Figure 2-5 with Figure 2-4 shows that combining the information that $\beta > 1.5$ with the data allows a much more precise assessment of B10. If the prior information is reliable, the impact on the inference can be substantial and important to exploit.

Combining Data from Multiple Sources

Here we consider an example of a more general situation in which the aim is to estimate the reliability of a motor component as a function of time. The example includes the following two assumptions: (1) the true reliability as a function of time can be represented as a member of a family of cumulative distribution functions indexed by a single parameter θ, which is the mean time to failure for each member of this family of distribution functions; and (2) we have useful information about θ from two experts on this component, three computer simulations, and five sets of data from physical experiments. How might these three disparate sources of information be combined to provide the analyst with both an estimate of θ and estimates of the uncertainty of our estimate?

Expert A believes that θ follows a normal distribution with mean 80.0 and standard deviation 4.0, while expert B believes that it follows a normal distribution but with mean 73.0 and standard deviation 4.0. Three computer simulations have been used to simulate the functioning of the motor component. The first simulation shows that estimates of θ are centered at 78.0 with standard deviation of 6.3, the second shows estimates of θ centered at 69.0 with standard deviation of 10.8, and the third shows estimates of θ centered at 67.0 with standard deviation of 6.5. Five types of developmental testing have been carried out on five sets of motors. For each set of motors, the means and standard deviations of the failure times were observed as follows:

	Mean	Standard Deviation
Test 1	87.0	5.0
Test 2	83.0	3.5
Test 3	67.0	3.0
Test 4	77.0	4.0
Test 5	70.0	5.0

Classically, these various sources of information would be joined using a linear combination of the separate estimates weighted inversely proportionally to their variances (i.e., the square of their standard deviations). (There is a further complication if the estimates are not independent.) In this approach, the computer simulations would be considered subject to between-simulation variance, as well as the within-simulation variance indicated above, which would be estimated and then added to each simulation's standard variance in calculating the optimal linear combination.

An alternative way of combining this information is through use of Bayesian prediction. Using Bayes' Theorem, prior probabilities are updated to posterior probabilities through use of the likelihood function, as in the above example on dichotomous outcomes where the likelihood was modeled using the binomial distribution. The prior is determined using the three simulations and the two experts, and the likelihood is based on the results from the five experiments. To determine the prior, as in the classical framework, one could use a linear combination of the five subjective information sources. One might provide each expert and simulation with weights that vary inversely according to their supplied variances, though a number of other approaches are also possible. One might also downweight estimates based on their distance from the estimated center of the five estimates (this could be iterated until convergence).

To build the likelihood from the experiments, we assume that the failure times have mean θ and standard deviations that we will estimate using the data (though combining information approaches to determine the standard deviations could also be used if there were relevant prior information). Using the assumption (based on expert judgment from previous experiments) that the estimates for θ from the five experiments have non-zero correlations ranging from 0.19 to 0.90, the five experiments support the model that individual failure times are normally distributed with mean 78.4 and standard deviation of 1.9.

The prior and the likelihood, using Bayes' Theorem, can then be used to produce the posterior distribution, which would now reflect the information from the experts, the simulations, and the developmental test results.

A number of assumptions were made to arrive at the final result, and at each stage sensitivity analyses should be used to assess the impact of divergences from these assumptions. For example, the assumption of normality is unlikely to be satisfied for failure times, but this discrepancy can be addressed by a number of modifications to the above procedure, such as transforming the data to enhance the fit to normality. Any assumptions not

A Treatment of Separate Failure Modes

Information from developmental testing can be used to make operational test evaluation more efficient when there are separate failure modes with varying failure characteristics. ATEC combines information from engineering judgment, analysis of data from developmental and other tests, training exercises, modeling and simulation, knowledge of redesign activities that occur after developmental testing, and other sources to create an operational test that will expose these failure modes. Through analysis of this information, situations can be identified, with associated prior probabilities, that indicate which of the failure modes in the developmental test remain active in the operational test. (In a less simplistic situation, one would, of course, be concerned with failure modes appearing in operational testing that did not appear in the developmental test.) The operational test data can then be used to update the estimated probabilities of these situations. This method is particularly helpful when trying to assess the properties of a large number of failure modes that either are statistically dependent or have, individually, low failure rates.

In principle, the computation is straightforward. In practice, however, a considerable level of expertise is needed to develop suitable prior information and combine it appropriately with experimental data. The following simplistic example demonstrates one approach.

During developmental testing a vehicle has exhibited two critical failure modes, mode 1 and mode 2. Both involve components with relatively mature designs, so infant mortality is not present. The vehicles have experienced relatively low usage, so wearout is not likely. For these reasons, or perhaps because the failures are due to external stressors exceeding a certain limit, it is assumed that each mode exhibits exponentially distributed times to failure. However, the failure rates, R_1 and R_2, are not known. Therefore we need to make statistically supportable statements about three performance measures when the system enters operational testing after modifications based on developmental testing:

- Λ_i = vehicle (total) failure rate per mile due to mode i in operational use,
- $MDTF_i$ = mean distance to failure due to mode i, and

EXAMPLES OF COMBINING INFORMATION 37

- Rel(m) = m-mile reliability = probability that a vehicle will survive m miles without failure.

Using engineering judgment and the results of developmental testing, it is assumed that we are able to consider four possible different situations:

- S_0 = no failure modes remain after developmental testing is concluded,
- S_1 = only failure mode 1 remains after developmental testing,
- S_2 = only failure mode 2 remains after developmental testing, and
- S_3 = both failure modes 1 and 2 remain after developmental testing.

We assume that we are comfortable in assessing a priori probabilities p_0, p_1, p_2, and p_3 respectively for these situations, and our uncertainty about the failure modes and the associated $MDTF$ can be expressed by assessing the expected value E($MDTF_i$) and standard deviation STDEV($MDTF_i$) for each mode.

Note that under this framework, the mean distance to failure is random, since it is unknown. We can update a prior distribution about the mean distance to failure, using operational test data, to arrive at a posterior distribution. This posterior distribution will itself have a mean, the expected mean distance to failure, and a standard deviation.

Now suppose that, after an exposure of t total vehicle miles in the operational test, n_1 failures of type 1 and n_2 failures of type 2 are observed (where n_1 and n_2 can be 0). Appendix B shows the development and specific equations that allow calculation of the three performance measures, as well as their uncertainty, expressed by their posterior standard deviations. For example, suppose expert information based on developmental testing and other activities provides us with the following prior values:

E($MDTF_1$) = 2,500; STDEV($MDTF_1$) = 2,000;
E($MDTF_2$) = 3,000; STDEV($MDTF_2$) = 3,500;
and
E($MDTF_0$) = 100,000; STDEV($MDTF_0$) = 0;

where the 100,000 mile (certain) $MDTF$ value reflects a practical assessment of the situation "no failure modes remaining." Using scenario probabilities p_0 = .005, p_1 = .10, p_2 = .15, p_3 = .745, Table 2-3 shows various performance measures for three potential values of (n_1, n_2) failures in t = 20,000 total exposure miles.

TABLE 2-3   Three Potential Values for the Number of Failures of Two Types Observed in 20,000 Miles, and the Resulting Impact on Reliability Estimates and Their Uncertainty

(n_1, n_2)	(0,0)	(0,1)	(1,1)
Posterior E(Λ)×1,000	.145	.187	.321
Posterior STDEV(Λ)×1,000	.102	.105	.111
Posterior E(*MDTF*)	17,073	7,744	3,543
Posterior STDEV(*MDTF*)	26,360	6,417	1,412
Posterior E(Rel(1,000))	.868	.834	.730
Posterior STDEV(Rel(1,000))	.085	.083	.079

This example reflects the use of weak prior information, in that STDEV(*MDTF*$_1$) and STDEV(*MDTF*$_2$) are about as large as their respective mean values. Therefore, the reported performance measures are relatively objective in that they depend mostly upon the operational test results. It is also possible to compute posterior probabilities for the four situations (see Appendix B) that show the same relative insensitivity to prior assessments.

This general approach can be extended to account for more complex situations, as in the following example. A system has 40 type A vehicles and 30 of type B. A developmental test has been run with miles of operation per vehicle ranging from 1,000 to 15,000 miles, with 10 failure modes discovered at various mileages. An operational test is then run with 24 vehicles, all of type A, with miles of operation now ranging from 500 to 2,000 miles. Four of the original 10 failure modes are observed, occurring at varying mileages but with a higher rate than in the developmental test. In addition, a failure mode is seen that was not present in the developmental test. The operational test is set in three different environments of use, and the developmental test has been exclusively in a fourth environment of use, a test track.

Although this approach to combining information from developmental and operational testing is a tempting means to increase the efficiency of operational test results, a number of potential difficulties remain. To the extent that an analyst must speculate about possible situations that have not been realized, an assessment of their probabilities may be more vulnerable to cognitive biases than the better understood assessment of distributions of more intrinsically engineering- or physically based parameters. In addition, the analysis necessary for the more complex combinations of failure modes implied in the more realistic example above will require expertise not necessarily resident at the test agency. Not inherently suitable for encapsulation in manuals or training courses, the methodology would require nonstandard certification for each use.

On the other hand, sensitivity analysis with respect to prior assessments can be readily performed using simple spreadsheet software models. Moreover, inferences made about performance measures are couched in language appropriate for decision making.

In summary, inferences about the number of failure modes that have been fixed prior to OT, the number of new failure modes that OT has introduced, and related problems can be addressed using combining information techniques. These techniques are strongly dependent on assumptions, and therefore their proper application requires the use of sensitivity analyses to determine dependence on various assumptions.

3

Combining Information in Practice

The previous chapter presented a number of examples of the use of techniques to combine information. In this chapter we discuss some considerations when implementing these techniques and the complications that often accompany analyses of operational test data in defense and related industrial applications.

The panel notes that, while the operational evaluation of the Stryker/SBCT is a large and extremely complex problem, this degree of complexity is not unique within the DoD or other government agencies such as the Department of Energy (DOE). Los Alamos National Laboratory (LANL), for example, must evaluate the weapons in the aging nuclear stockpile and certify their safety, reliability, and performance even though the live test data that have traditionally been used for this evaluation can no longer be collected.

For its evaluation of the nuclear weapons stockpile, the Department of Energy is developing approaches that employ formal methods for using expertise and combining information. Although live, full-system test data are no longer available, there is a great deal of relevant information—including results from computer simulations, historical test data, subsystem tests, and expert judgment—available through a large and multidisciplinary community that includes engineers, physicists, materials scientists, statisticians, and computer scientists. Traditional reliability demonstrations would be very difficult, and traditional statistical methods must be significantly expanded to include the representational methods discussed above and the information-combining methods discussed here and in Chapter 2. An example of how these methods might be applied to a large, complex system is given in Appendix C.

COMBINING INFORMATION TO ASSESS SUITABILITY, SURVIVABILITY, AND EFFECTIVENESS

The operational test for Stryker is intended to assess a large number of performance criteria. In the system evaluation plan (SEP) for Stryker the measures of performance and effectiveness (MOPs/MOEs) are grouped into three areas: suitability, effectiveness, and survivability. Suitability encompasses issues such as transportability, maintainability, availability, and supportability. Measures under this broad heading are often not situation dependent, and so combining information from the operational test with that from training, developmental tests, and perhaps testing and field use of similar systems can often be relatively straightforward. For example, all instances in which Stryker is found to be transportable on a C-130 aircraft, whether from a training exercise or in developmental or operational testing, provide valid information about transportability. The various methods described above for combining information for use in assessing reliability (and other related methods) can be effectively applied in this area.

Measures of survivability and effectiveness, on the other hand, are typically situation dependent. Information from operational training missions (such as raids and perimeter defense) is not easily combined with information from operational test missions because of the many differences between training and test operational situations. The approach used most often to combine information about survivability and effectiveness is the combination of information from operational tests, conducted by ATEC, and modeling and simulation efforts, such as those obtained by the U.S. Army Training and Doctrine Command's (TRADOC) Analysis Command. Methods of combining information useful for modeling measures of system survivability and effectiveness are likely to require relatively specialized models of system performance, which are typically achieved through modeling and simulation.

The combination of information from tests and simulations is already standard DoD procedure. Modeling and simulation results play a part in designing operational tests, and the results of operational tests are used to refine and improve modeling and simulation programs through a model-test-model approach. This existing DoD activity is an example of the wide range of methods subsumed under the rubric of combining information.

The Stryker operational test will provide quantitative information that can be used in subsequent modeling and simulation efforts (though such efforts will likely not be used for the operational evaluation of Stryker). This information includes detailed performance measures such as detection times, detection probabilities, time between rounds fired, and probability of surviving direct hits, which can be used as direct inputs to detailed simulations. The operational test can also provide data on sample attrition rates that can be used as input to aggregated models. In either case, the simulations and models could then be used to augment the limited number of situations considered in the operational test by simulating other operational situations to provide a larger base of information for evaluating the survivability and effectiveness of Stryker and the SBCT.

There is relatively new, relevant statistical research on combining information from experimental systems with that from computer models (see, e.g., Reese et al., 2000). One important, and challenging, step in carrying out this type of information-combining is to assess the variability and uncertainty in the output of the computer models that result from poor or insufficient inputs. The uniqueness of each application and the fact that the research is still evolving prevent our making any general statements about approaches that ATEC should take along these lines.

ISSUES IN COMBINING INFORMATION FOR RELIABILITY ASSESSMENT

Reliability is typically defined in textbooks as the probability of survival (or operation without failure) for a given mission time and under specified conditions. A more practical definition would identify and carefully characterize encountered conditions, recognizing that most systems have to operate in a complicated, dynamic environment.

Customers generally desire information or assurance about the reliability of a system or product before they decide whether to purchase it and for what price. Manufacturers, for their part, need to assess a product's reliability before it is released in order to reduce the risk of serious field reliability problems and warranty costs. A purely empirical reliability demonstration typically follows the significance testing framework described in the NRC's 1998 report (pp. 88–91) and exemplified in DoD documents such as MIL-STD-690C (*Failure Rate Sampling Plans and Procedures*), MIL-STD-781C

(*Reliability Design Qualifications and Production Acceptance Tests: Exponential Distribution*), and MIL-HDBK-108 (*Sampling Procedures and Tables for Life and Reliability Testing—Based on Exponential Distribution*).

The fundamental ideas behind reliability demonstration testing are straightforward; an example in this instance is the specification that mean time to failure (MTTF) for a Stryker vehicle should be at least 10,000 miles. In order to demonstrate that this specification has been met, it is necessary to have a test that results in a lower confidence bound on MTTF that exceeds the specification. A minimum sample-size plan to make such a demonstration may have appeal, but to have a reasonable probability of successful demonstration, the actual MTTF would have to be much larger than 10,000 miles. Thus, under the simplifying assumption of an exponential failure time distribution having only one unknown parameter, a demonstration at the 95 percent level of confidence would require testing three units for 10,000 miles and having no failures (see, for example, equation (10.01) in Meeker and Escobar, 1998). If the true MTTF is 15,000 miles, the probability of a successful demonstration (i.e., no failures) is only $\exp(-1/1.5)^3 = 0.135$. If the true MTTF is 30,000 miles, the probability of successful demonstration increases to $\exp(-1/3)^3 = 0.368$, which is still not very high.

Although larger sample sizes can provide higher probabilities of success by allowing for a small number of failures during the test, these sample sizes can increase dramatically when one must estimate two parameters (e.g., fitting a more realistic Weibull distribution with an unknown shape parameter). Thus, although these methods of reliability demonstration are useful for testing materials or components, unless the actual reliability is very much greater than the specification, they are generally impractical for large, expensive systems, because large sample sizes or unrealistically long tests are required.

The previous illustration should make it clear that unless the true reliability of a system is overwhelmingly high, one will need very large amounts of reliability data to achieve the desired goals of reliability *demonstration* with some confidence. A number of information and data sources for both quantitative and qualitative information are available for such an evaluation of the Stryker/SBCT. The major sources include operational testing, developmental and technical testing, contractor testing, data from previous tests of similar systems, training exercises, experience of foreign armies with variants of the Stryker (though these systems are not very similar, which would severely limit the value of this information), engineering judgment, military judgment, and modeling and simulation.

The goal is an assessment, referred to as a reliability *assurance*, that is not as rigorous a confirmation as a reliability demonstration but that can still provide sufficient information on which to base a decision on promotion to full-rate production. In this approach, data are combined from a variety of sources, and the inference, as a result, is more model-based than in a reliability demonstration.

The following discussion addresses the use of these sources and considers specific formal methods.

Use of Military Judgment

It is always encouraging when statistical analysis of data harmonizes with the judgment obtained from insight, intuition, and experience. Of course, one should also consider how each may influence the other. Does the data analysis trigger the harmonizing after the fact? Would other results have led to other harmonies? It is much more convincing if evaluators and those providing other information write down their analysis results or insights and intuitions before comparing them for validation. Unfortunately, even in this case, minor differences will often be explained away if there is pressure for a certain interpretation of the results.

Combining Test Data

Operational testing for the Stryker will involve many vehicles over relatively short exposure periods. Unless one is analyzing failure modes with lifetimes that are reasonably described by an exponential distribution, the summary experience over these many short exposure periods is not equivalent to the summary experience of a few vehicles over long exposure periods. This is the case even when the total exposure time for both sets of vehicles is the same. Data from such longer exposure periods may well be available from developmental testing but only for a few vehicles.

Combining Test Data: Exponential Models

The assumption that individual components and replaceable units (not repairable systems) have lifetimes that follow an exponential waiting time distribution may be reasonable in situations where the failures are mostly due to external stressors exceeding a certain limit. Such a limit characterizes the vulnerability of the fleet of vehicles. However, before employing an exponential lifetime analysis, it should be confirmed that this vulnerability is not affected by aging. Such a confirmation almost always will involve the expert judgment of those who perform postmortem analyses of component and replaceable unit failures. The judgment to use an exponential distribution is an implicit form of combining information, since one is using expert opinion to stipulate a specific distributional form, in this case that the shape parameter in a Weibull model is equal to 1.

When an exponential failure time model is appropriate, the combining of data from two or more sources is fairly straightforward, provided the failure rates are roughly the same. The number of failures is combined into one overall count N and the exposure times into one overall total exposure time T, and the analysis is performed using these two entities, with N/T being the maximum likelihood estimate of the failure rate. Here the two or more data sources can be operational, developmental, training, or other exposure tests or exercises, or the data may be obtained from subsystem experiences. In the latter case, the analysis is performed as though failures from all of these subsystems can be treated alike, as a common failure mode.

It is essential to also compute individual failure rates together with their uncertainties to judge the assumption of homogeneity. Such a judgment can be informal (e.g., using a graphical technique) or formal (e.g., using significance tests). When applied to small data sets, such judgments tend to be liberal in that homogeneity will not be easily rejected unless the differences are sufficiently large. This will lead to pooling of data with minor differences, and the mixed populations will exhibit somewhat higher variability characteristics than each contributing population. Such pooling of inhomogeneous exponential data gives the impression that the underlying failure phenomenon has a decreasing failure rate as opposed to the constant rate characterizing the exponential model (see Proschan, 1963). The result will be a better understanding of a mixed population instead of a more vague perception of many individual populations.

When the failure rates under different exposure regimes (e.g., the operational test and developmental test) or for different categories of subsystems show significant variations, it may still be possible to determine whether those variations are due primarily to a single factor. For example, failure rates during developmental testing may differ from the rates under operational testing, but for a particular group of failure modes the ratios of failure rates under the operational test to those under the developmental test might be roughly constant. (This is the approach taken in Samaniego et al., 2001.) If this constant were, for example, 2, it would mean that operational test failures occur at roughly twice the rate of developmental test failures. An explanation might be that the external stressors (e.g., rugged terrain, wet weather, or rougher driving styles) in the operational test exceed the vulnerability limits approximately twice as often. For example, ball bearings can be damaged by sufficient shocks caused by rough terrain or unskilled driving (e.g., hitting a curb with the wheel). Even though bearings eventually wear out, a postmortem analysis of failures may be able to distinguish (e.g., by comparing the defective bearing with other bearings on the same vehicle) between the strong shock casualties and those that come from normal wear. This is another example of combining information obtained from engineering judgment used in conjunction with actual data. (Note that although this example is presented, for ease of explication, in the context of exponential lifetime analysis, it applies as well to other lifetime models.)

If data from several previous systems are available during the developmental and operational tests, and if one finds that for specific components a failure rate during the operational test is roughly a certain multiple of the corresponding failure rate under the developmental test, then such a factor could be used to analyze the data for a current system for the same type of component in a combined fashion. The broader the prior experience over which this factor appears to be constant, the more confidence one can have in the use of such a factor for the situation at hand.

This kind of analysis requires the foresight to have collected and archived data for easy retrieval. Unfortunately that is usually not the case in industry or in defense acquisition, because it is hard to convince the financial decision makers to spend money on projects that are not immediately useful and may pay off only in the future, for a different program, after several such systems have been built and tested. The utility of establishing and maintaining a data archive is discussed in Chapter 4.

The common factor approach can be extended to more complex and flexible regression models where (often the logarithm of) the failure rate is modeled as a linear combination of known factors that may influence the failure rate in some form. Such factors could identify the environmental exposure conditions or different mission scenarios during which failures occurred. As mentioned above, the exponential distribution is appropriate when failures occur due to random external shocks. Such regression models, when they do not involve too many independent parameters, can lead to strong pooling of information, i.e., to a great reduction in estimation uncertainty when compared to separate analyses based on data for each factor combination.

If individual failure rates appear to be sufficiently different and combining data is not an option, this finding in itself is a form of combining information. Namely, more is learned from the collective of individual pieces of information than from each piece by itself; in this case it is learned that they are different, and the source of that difference can be investigated. This comment applies not just in the exponential lifetime context but in all others as well.

Even in this situation different failure rates can be treated as random effects. By estimating the variability of these rates from the individual sources, pronouncements can be made about the collective of such rates if they can be reasonably viewed as a random collection from some population. Here there is a trade-off between a larger data collective and a somewhat more uncertainly defined population, i.e., between a relatively large variance for the random effects and a relatively small variance.

Combining Test Data: Weibull Models

A popular extension of the exponential model is the Weibull model, which not only describes the lifetimes of components and replaceable units that fail due to external causes, but also provides a framework for lifetimes that arise from wear-out failures or infant mortality. Wearout failures are quite common for mechanical systems, gears, axles, bearings, clutches, and brakes. Infant mortality failures arise in some electronic components and subsystems.

These two kinds of failure can be effectively represented with a Weibull distribution, which is intrinsically identified by two parameters, the characteristic life η (acting as a scale parameter) and the shape parameter β, governing the skewness of the distribution. Symbolically, we have:

$$f(t) = \frac{\beta}{\eta}\left(\frac{t}{\eta}\right)^{\beta-1} e^{-\left(\frac{t}{\eta}\right)^{\beta}}.$$

On a logarithmic scale for the lifetimes this distribution becomes a location-scale family with location parameter $u = \log(\eta)$ and scale parameter $b = 1/\beta$. When $\beta = 1$, the Weibull distribution yields the exponential distribution as a special case. Situations with $\beta > 1$ are appropriate for describing wearout and other phenomena (and $\beta < 1$ for infant mortality).

As mentioned previously, estimating both Weibull parameters η and β entails an additional uncertainty in the estimation process and therefore has more stringent data requirements. Here the case for combining information becomes even stronger than in the exponential situation. If the shape parameter β is approximately known from previous experience, the Weibull lifetime data individual values X_i can be transformed via $X_i \rightarrow X_i^\beta$ into exponentially distributed data, and all the methods discussed above carry over. If working with a known shape parameter is problematic, several values can be used in a sensitivity analysis, and, depending on the application, one of these can be used as a conservative choice. For example, when it is clear that the system is subject to wearout, $\beta = 1$ can be used as a lower bound on β. For some situations this will yield conservative results (see, for example, Section 10.6 in Meeker and Escobar, 1998).

When β must be estimated as well, data can be combined using the assumption that the two sets have the same shape parameter but possibly different η's (the assumption of common shape parameter should be checked formally through tests or informally through graphical tools). In this fashion the uncertainty in estimating β will be greatly reduced. Considering the logtransform of Weibull lifetime data, this is essentially analogous to pooling variances, as discussed earlier.

Further methods for combining Weibull data are similar to those described for the exponential model, culminating in a linear regression model that treats $\log(\eta)$ as a linear function of various known factors that vary across all lifetime data that are intended to be used in the combination effort. Here again, the underlying assumption that only η varies and not β must be assessed.

For a sequence of failures of repairable systems, the distribution of the times between failures of a particular system component often depends not only on the nature of the repair or component replacement but also on the general state of the system, which, in turn, may also involve the specifics of maintenance actions carried out over time. Even so, it may be possible to model component lifetime distributions as a function of related explanatory variables.

An alternative method for modeling reliability data from repairable systems is to use a stochastic process model for events in time. Such a process can be characterized by representing the failure intensity as a function of variables such as the age of the system, the environment in which the system operates, and other changes as they occur over that system life. Such models are especially useful when modeling system reliability and availability and when tracking costs of repair and operation. An extensive treatment of the relevant issues can be found in Ascher and Feingold (1984), Meeker and Escobar (1998, Chapter 16), and Nelson (2003).

Industrial Experience and Stress Testing for Reliability Assurance

Increased market competition has resulted in widespread cost cutting, which increases the likelihood of reliability problems by reducing the ability to build in traditionally large factors of safety. These issues have driven some manufacturers to use new methods of manufacturing and reliability modeling, assessment, and improvement, taking advantage of new technologies. Examples include monolithic (as opposed to built-up) structures, accelerated testing, robust design, computer modeling, importance sampling in fault tree analyses, increasing reliability through redundant system design, probabilistic design, and structured programs for design for reliability, such as design for six sigma.

Reliability practices and procedures differ from industry to industry and from company to company within an industry, and often remain proprietary, especially with respect to the development of models that can be used to more effectively predict reliability without having to do expensive physical testing.

In a reliability assurance program, the overall goal is system reliability, generally determined by past product experience and benchmarking against best-in-the-industry competitors or by a marketing need to have a warranty period of a certain length of time. Metrics used include percent of returns within the warranty period or average warranty costs per unit sold. Failure modes and effects analysis (FMEA) and reliability block diagrams are used to quantify the relationships between the system, subsystems, components, interfaces, and potential environmental effects; these quantified relationships are referred to as the reliability model.

To meet the overall reliability goal, a reliability budget is developed to allocate reliability goals to different subsystems. For example, in the aircraft industry a 10^{-9} risk for a critical subsystem failure is often used as the targeted goal to maintain the industry "standard" of one critical aircraft failure in about 10^6 to 10^7 flights and the assumption that there are about 100 such subsystems to monitor. However, such 10^{-9} risk goals are usually established through modeling, since real experience on this order is not attainable. Furthermore, such risk levels are often not accompanied by confidence bounds that reflect the uncertainty of any data utilized in such an analysis. This is due partly to the difficulty of achieving even an estimated 10^{-9} risk goal and also to the problem of reconciling two such disparate risks, namely 10^{-9} and the 5 percent chance of missing the target with the confidence bound. Even if the reliability for aircraft as high as $1-10^{-6}$ or $1-10^{-7}$ per flight is the currently tolerated level, there are industrywide efforts under way to significantly increase this reliability level because of the anticipated growth in airline travel. At a constant accident rate the public acceptance of the resultant growth in the number of accidents is not a given. Each industry has its own considerations and sensitivities in budgeting such subsystem reliabilities; for instance, major recalls in the automobile manufacturing industry are not uncommon and can be very costly.

Inputs to reliability models, including associated uncertainties, need to be determined. Assuming the same or similar environmental conditions, previous experience with particular materials and components can be used directly; examples include experiences codified in MIL-HDBK-5 and MIL-HDBK-17 (handbooks for metals and composite materials) through A- and B-allowables, with 95 percent lower confidence bounds on the 1 percent- and 10 percent-points of the strength distribution for a given material. Because of the wide acceptance of allocating reliability as a concept in structural design, they have found use in nonstructural arenas as well.

Computer modeling, along with appropriate physical testing to verify the accuracy of the model, can often be used to provide needed information on component reliability. The multitude of factors involved and the occasionally high cost of simulation runs has led to an entire subfield of design and analysis of computer experiments.

Adjustments are made to critical components in each subsystem in order to meet subsystem reliability goals. Testing of a small number of prototype subsystem units at higher than typical use conditions can be done in order to discover weaknesses. These tests represent a kind of accelerating testing, which can take various forms, some of which are described in McLean (2000). When new failure modes or weaknesses are discovered, design changes should be considered, albeit with the understanding that failure modes generated in the test might never occur in actual operation and that money spent on design changes might therefore be wasted. Another risk is that some failure modes revealed by the accelerated testing could mask other failure modes that might not appear during the accelerated testing and thus remain undetected and uncorrected.

After the complete system is assembled, it may be necessary to conduct durability tests for certain parts of it. In some cases, this is done economically by testing a small number of systems or nearly complete systems using continuous-use testing or rapid cycling, as appropriate. Separate tests may have to be conducted to excite different failure modes; for example, in automobile engine testing there is a standard test using a continuous run protocol and another that uses a start-stop-start protocol. While it is feasible and effective to use up-front testing of components and subsystems to assess their reliability characteristics, the same is not usually true for major systems whose reliability goals and costs are very high.

Methods of strenuous testing of early production units are often employed to discover reliability problems before large quantities of product have been shipped. For example, manufacturers of washing machines may have an arrangement with laundromats, and automobile manufacturers may track fleets of early production units with friendly customers. In both cases, the manufacturers track warranty returns to learn as early as possible about problems so that they can be corrected.

Another example is the staggered entry into service of new aircraft for which the timing and location of first fatigue cracks or corrosion are carefully recorded, so that succeeding aircraft of the same type can be examined and maintained more aggressively; thus past experience is used to indicate which areas to monitor for cracks and corrosion. For such an approach to be effective, proper maintenance schedules must be followed, incorporating any knowledge of cracks and corrosion or other wear of materials, while also allowing for the probability of nondetection during an inspection.

When sufficient information is not available from other sources, physical testing (e.g., accelerated life or durability tests) may have to be conducted. If adequate physical testing cannot be done, then uncertainties may be addressed through the use of design safety factors, although this practice lacks scientific rigor. Usually such tests involve samples whose size is constrained by costs, and the possible variability underlying the test (because of the small sample size) is absorbed or accounted for by increasing the reliability by a factor (derived mainly from engineering experience) that is considered acceptable.

While use of design safety factors is an example of combining information (test results with engineering judgment or industry culture), such factors are difficult to rely on since they have no probabilistic interpretation. Sometimes they are intended to implicitly account for the "unknown unknowns" (or UNKUNK) and appear to offer insurance for unforeseen contingencies. However, failure to examine the degree to which this is true empirically does not support this use or interpretation.

Field experience typically validates the use of design safety factors, albeit conservatively, because systems designed according to safety factors often satisfy their reliability requirements. Furthermore, this design process has also had additional benefits. For example, there have been incidents where aircraft were stressed far beyond design loads and survived with just the wings bent out of shape, and in one case this led to improved aerodynamic wing properties. Such "success stories" have led to a strong resistance to change among some members of the engineering design community.

But while safety factors may be cheap during design, they often increase both the purchase cost and the costs that accrue during the lifetime of the product. In the aircraft industry, limited checks on the possibility of overdesign are carried out when a new wing design is statically loaded until it breaks, the aim being that the strength of the wing not exceed the design value by more than is necessary. Similar cyclic dynamic tests examine a new aircraft frame for fatigue failures. Because such factors typically have no known associated reliability, a major analysis of them based on analytical probabilistic design models and experience in the field could have long-range benefits.

4

Prerequisites for Combining Information

The development and implementation of techniques for combining information, whether in design or in evaluation, are often sophisticated activities. What may conceptually seem to be relatively straightforward applications often require original thought, nontrivial modification of existing techniques, and software development. But the use of methods for combining information can be made easier if the appropriate methodological and logistic frameworks are in place. This chapter discusses several key steps that should be taken to establish these frameworks: broader definitions of data so that nontest data (e.g., expert judgment and computer models) can be formally and correctly included in analyses; development of test data archives so that what is learned about a system continues to be of use to future evaluators; use of graphical representations of complex systems to aid in the understanding of overall reliability and performance; and use of formal statistical methods for information combination.

This chapter also identifies the statistical capabilities required to implement such strategies. There is no clear evidence that the service test agencies have these capabilities in place today, and so if they find the advantages presented here compelling, it will be necessary for them, with help from higher-level officials within the services, to acquire the capabilities described in this chapter.

NEED FOR A BROADER DEFINITION OF DATA

When performing an assessment of a complex system, the most commonly used data are test data, whether from operational, developmental, or contractor tests. Other sources of information about the system include training exercises, field use, computer models and simulation, and military and engineering judgment.

Figure 4-1 is a schematic diagram of a system and its available data sources. If resources were available, it would be desirable to collect test data on every part of the system and to perform system tests under a variety of conditions. For large and complex systems, however, that is seldom possible, and so the assessment often resembles that shown in Figure 4-1, where some parts are not tested at all, some have computer modeling and simulation data, some have historical data, others have test data, and some have multiple sources of data.

The challenges in methods for combining information are to (1) represent the system under test in a way that all of the stakeholders can understand (in Figure 4-1, a fault tree is used, one of many useful representational schemes); (2) collect data (broadly defined) to assess the system and map them onto the representation; and (3) perform appropriate statistical analyses to combine the available information into estimates of the metrics of interest. All of these steps are performed in some way by ATEC's current operational evaluation; this chapter provides suggestions for additional capabilities. For example, the graphical representation of Figure 4-1 could be used to facilitate understanding of the system evaluation plan data source matrix, to suggest areas where data are (or will be) missing and where data combination is possible, and to provide a structure for test planning.

It is important to acknowledge and account for possible weaknesses in different kinds of data. The use of nontest data for evaluation can be contentious, although it is done routinely. Military, engineering, and statistical judgments are required to design test plans and interpret data; and computer modeling and simulation are applied to test data collected under certain scenarios to extrapolate the scope of their validity to other scenarios or to larger fighting units. Methodological contention arises when attempts are made to use military judgment or computer modeling and simulation results formally as data, instead of using them only to inform design, modeling, or interpretation.

The use of expert judgments, in particular, is especially vulnerable to inappropriate application due to procedural or cognitive biases. For

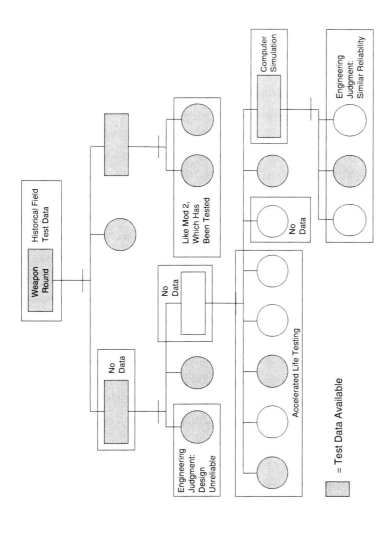

FIGURE 4-1 System with heterogeneous data sources.

example, suppose that test data are not available for a particular component but that engineering judgment considers the system design "unreliable" (Figure 4-1). Methods have been proposed (Meyer and Booker, 2001) to formally elicit and quantify engineering judgment for inclusion in statistical calculations, and there is a growing body of literature by statisticians, decision analysts, social scientists, and cognitive psychologists, developed over the past two decades, describing methods for eliciting and using expert judgment. Using information based on expert judgment requires considerable care, explicit documentation, and careful sensitivity analysis. With the recognition that all statistical analyses depend, to some degree, on subjective judgment (Berger and Berry, 1988) comes the obligation to ensure that such judgments are made in a rational and defensible manner.

It is well established that the major barrier to successful elicitation is the presence of biases inherent in the process used to evoke expert responses (see, for example, the pioneering work of Tversky, Slovic, and Kahneman, 1985). These biases are often characterized as cognitive or motivational, and attributed to a variety of sources, including: intrinsic cognitive failures, the instrument used to elicit responses, the social or institutional setting within which the expert operates, and the response mode.

Cognitive biases are evident in effects such as *anchoring*, the tendency not to adjust from a first response even after receiving information contrary to the position; *availability*, the elicitation of event probabilities or other values based on what readily comes to mind; *conservatism*, a reluctance or inability to draw inferences agreeing with those that would be obtained using Bayes' rule; and *underestimation*, an understatement of the uncertainty of an assessment. Motivational biases include *group think*, whereby experts tend to slant their assessments to what they perceive to be a consensus; and *misinterpretation*, in which the method or instrument of elicitation affects the expert's responses (as when, for example, the framing of a question cues the expert to provide a preferred response).

The test and evaluation environment contains strong institutional incentives and is therefore possibly subject to equally pervasive motivation biases on the part of experts asked to provide their judgment. These experts can be specifically trained in methods to avoid or mitigate an array of cognitive biases, and the elicitations themselves can be structured to minimize the effects of bias. A growing literature of methods addresses these issues. In particular, Meyer and Booker (2001) and Booker and McNamara (2003) provide exemplary guides to such ameliorative methods as indirect probability assessment, use of documented processes for elicitation, expert identification, motivation and training, modes of communication, and appropriate framing. If expert judgment is used in methods for combining information, it is extremely important that these or similar techniques be used, especially when arriving at prior distributions for critical parameters, such as failure rates.

Some industrial organizations have become comfortable using such techniques, while being aware of and adjusting for potential biases, in high-profile, politically sensitive analyses. For example, General Motors Corporation reports on its ability to assess technical success probabilities in Bordley (1998) and has used panels of over 40 experts to develop cumulative prior probability distributions for the improvement of fuel economy by using a novel powertrain concept.

Computer modeling and simulation, which can be thought of as combining the original data with the knowledge incorporated in the model, can also provide a cost-effective way of expanding the use of the data. The appropriate use of computer modeling and simulation methods depends crucially on the trustworthiness of the models in transporting data to other scenarios. Although simulation can generate a large amount of new data, it is a serious mistake to combine these generated data directly with the original data to increase the sample size. Instead, more sophisticated statistical methods (e.g., as described by Reese et al., 2000) should be employed.

NEED FOR A TEST DATA ARCHIVE

Given the wide variety of data sources available when performing a system assessment, a mechanism should be developed to archive the data and make them available for current and future assessments. At present, such data are not saved in a readily accessible database along with contextual information. This is true even for previous development stages of a system. Once a system has been fielded, the absence of rigorous information on system performance greatly limits the effectiveness of feedback loops relating performance in the field to performance during testing, feedback that could be very useful for improving system designs, the system development process, and operational and developmental test design.

A data archive of military system performance could be put to several uses that would assist in test design and system evaluation. In support of test design, data archiving can be used to:

- help set the requirements for the test design and develop the operational mission summary/mission profile (OMS/MP);
- determine the set of conditions and miniscenarios to be included in the developmental and operational tests;
- identify scenarios in which the new system is expected to perform better than previous systems (e.g., by providing information on how other systems performed in similar scenarios);
- similarly, identify scenarios in which the new system may perform poorly;
- identify factors that have an important impact on system performance;
- understand the factor levels that stress the system weakly, moderately, and severely; and
- determine adequate sample sizes through power calculations.

In support of system evaluation, data archiving can be used to provide information to support analysis of the validity of computer models and simulations used in test evaluation; support identification of appropriate statistical models for use in system evaluation; and support pooling and other forms of information combining. With the increasing development of statistical methods and models for information combining this last reason has become increasingly more compelling.

Data archiving can also contribute to improvement of defense system assessment by providing a means to better understand the differences between failure modes and failure frequencies in moving from developmental to operational testing and from operational testing to field use; understand the sources of system deficiencies identified in the field, which can then be used to guide design improvements; improve both developmental and operational testing and evaluation, e.g., by understanding how deficiencies identified in the field escaped detection in the developmental and operational tests; and estimate system and component residual lifetimes and life cycle costs.

The current lack of priority for data archiving, given the above advantages, suggests that the primary purpose of test data is to evaluate a system for promotion to the next stage of the milestone process of defense system development. Processes and techniques for combining data across acquisition stages either for a given system or across systems are not currently envisioned or well supported. However, such data, often acquired at enormous cost (e.g., operational tests can cost many millions of dollars), could and should be stored in an accessible form that would facilitate the above uses. Averaged over all defense systems in development, the cost of such an archive would be extremely small, but its value, as has been discovered in many industrial settings, could be substantial.

A test data archive would need to contain a rich set of variables to adequately represent the test environment, the system under test, and the performance of the system. Failure to initially include such a comprehensive set of variables should not be used as an argument for not getting started, since many of the potential benefits from such an archive could be derived from a subset of what is described here, with increasing detail added over time.

In order to accurately represent system performance, including the appearance of various failure modes and their associated failure frequencies, the circumstances of the test must be understood well enough that the test, training exercise, or field use can be effectively replicated, including the environment of use (e.g., weather, terrain, foliage, and time of day) and type of use (e.g., mission, intensity, and threat). This information is not easy to collect in controlled settings such as operational testing, and is considerably more difficult to collect in less controlled types of use, such as training exercises or field use. However, much in this direction can be accomplished. In addition, contextual information that might be relevant for an operational test might have little relevance in the developmental test, because often only particular components are under test.

While a system is under development, the system design is often under constant modification. Given the need, stated above, to be able to replicate a test event in the database, it is crucial to represent with fidelity the system that was in operation during the event so that proper inference is possible. Since modifications can and do occur during late-stage operational testing and after fielding, this is not only a concern for the developmental test. Even for systems produced at the same stage of development, knowledge of the order and location of manufacture can be useful to understanding why some prototype systems perform differently from others.

In addition to storing the length of time between system failures, it is also important to identify which hardware or software component malfunctioned; the maintenance (including repair) record of the system; the time of previous failures; the number of cycles of use between failures; the degree of failure; and any other variables that indicate the stresses and strains to which the system was subjected, such as speed and payload. It is also useful to include the environments and stresses to which individual system prototypes have been exposed historically (e.g., in transport, storage, and repeated on/off cycling), in order to support comprehensive failure mode analysis, especially if an apparent declining trend in system reliability appears. This sort of information is difficult to collect in less controlled settings; however, in many industries sensors have been attached to systems to collect much of the information automatically.

The information stored should be both quantitative and qualitative. The latter is important to include because the contextual information needed to help recreate the environment of use often includes qualitative information. To facilitate use across services, such an archive should make use of terminology common across services and, in its design and accessibility, should address classification issues.

With respect to the structure and function of the database, it should be able to track failures over time and identify systems that, while considerably different, have similar components. These needs argue for a database in which these linkages are facilitated. An analysis of similar data archives in industry would enable the DoD to build on existing processes and techniques.

The panel is pleased to note that there are defense databases that satisfy some of the above needs; the ATEC Distributed Data Archive and Retrieval System and several servicewide reliability or failure reporting databases are leading examples. However, those that the panel has seen support only a few of the potential benefits listed above, rather than the breadth, structure, and accessibility that we envision.

The marginal costs of data collection, input, and maintenance could be easily met through routine allocation of a small percentage of the development funds from every ACAT I program. The initial fixed costs for the Army might be funded by the Army Materiel Command and other related groups.

Finally, as mentioned in Chapter 5 for the Future Combat System, systems developed using evolutionary acquisition provide an additional argument for the establishment and use of a test (and field) data archive, since it is vital to link the performance of the system as it proceeds through the various stages of development. This test and field data archive could (1) assist in operational test design for the various stages of system development, (2) help in diagnosing sources of failure modes, and (3) assist in operational evaluation.

Recommendation: The Department of Defense should provide the funds to establish a test data archive that will be a prerequisite for combining information for test and evaluation of future systems.

REPRESENTATIONS

The fault tree represented in Figure 4-1 captures logically how the parts of the system under study interact. The same can be conveyed in reliability block diagrams. These and other classes of representations can be quite useful when assessing systems as large and complex as those evaluated by ATEC. For large, complex systems with heterogeneous data sources, representations of a system have several advantages: they set out a common language that all communities can use to discuss the problem; heterogeneous data sources can be explicitly located in the representation; and the representation provides an explicit mapping from the problem to the data to the metrics of interest.

If the system and its assessment are to be put in a decision context—for example, an overall assessment of system effectiveness and suitability supporting an acquisition decision—the fault trees and block diagrams may need to be embedded in a representation that supports these broader goals and connects the disparate and heterogeneous sources of data. Within the data archive one can use representations of the test environments to understand and compare variables such as the environment of use (weather, terrain, foliage, and time of day) and type of use (e.g., mission, intensity, and threat) across multiple tests.

It is important to develop a set of higher-level representations of the system under evaluation for use both within the data archive and more broadly in the system assessment. These representations, of necessity, change over time as the system and the context of the evaluation change. Standard reliability assessment methods focus on individual parts or simple groups of parts within a system. Assessing the overall reliability and performance of a complex system, however, involves understanding and integrating the reliabilities associated with the subsystems and parts, and this understanding and integrating are not always straightforward. Multiple and heterogeneous data types may exist, and the wider community that owns the system may not understand all the features and relationships that can affect system reliability. One way to illustrate all of the factors that characterize and impinge upon system reliability is by building qualitative graphical systems representations that can be migrated to graphical statistical models to assess reliability. For this reason, it is important that the information stored in a data archive be both quantitative and qualitative, as noted above.

Most groups developing complex systems do develop compartmentalized graphical representations of reliability. These representations may include reliability block diagrams; timelines, process diagrams, or Gantt/PERT charts dealing with mission schedule and risk; and engineering schematics of physical systems and subsystems. But none of these disparate representations capture all aspects or concerns of the integrated complex system. Moreover, since the system is likely under development with users, procurers, planners, managers, designers, manufacturers, testers, and evaluators spanning multiple organizations, geographical locations, and fields of expertise, these numerous, specialized, and compartmentalized representations foil attempts for the multiple groups to meaningfully discuss (or even understand) total system reliability or performance.

There are sets of methods and graphical representations that capture the full range of features and relationships that affect system reliability. For example, Leishman and McNamara (2002) employ ethnographic methods to elicit a model structure from the pertinent communities of experts involved in developing the system. The information on system reliability can initially be captured using "scratch nets" (Meyer and Paton, 2002) (i.e., simple diagrams that sketch out the important features of the system and its decision frame), which also allow a preliminary mapping of the key relationships between features. These scratch nets form the basis for more formalized representations called conceptual graphs (Sowa, 1984), which are a formal graphical language for representing logical relationships; they are used extensively in the artificial intelligence, information technology, and computer modeling communities. Similar to the less formal scratch nets, conceptual graphs use labeled nodes (which represent any entity, attribute, action, state, or event that can be described in natural language) and arcs (relationships) to map out logical relationships in a domain of knowledge.

The example in Figure 4-2 is a typical use of a conceptual graph to convey the meaning of natural language propositions within the context of a complex system. Generally, representations of complex systems are used to capture higher-level concepts, but the grounding of conceptual graphs in both natural language and formal logic also allows them to be used for expert judgment elicitation (and even potentially for text mining) to build

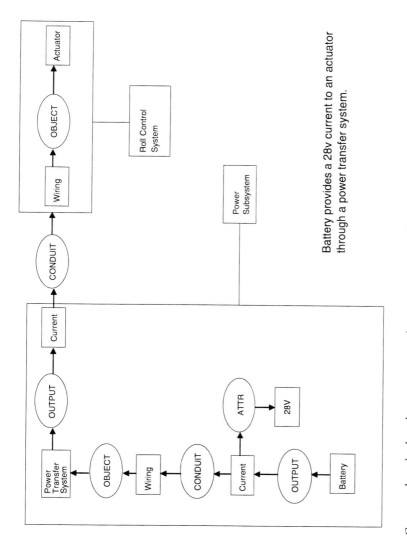

FIGURE 4-2 Conceptual graph adapted to represent complex system concepts and natural language about the system.

formal logic models that can then become formal mathematical and statistical models.

From the initial scratch nets, conceptual graphs are used to create an ontology (i.e., a representation of high-level concepts and main ideas relating to a particular problem domain) for the system. This ontology represents the major areas of the system and its decision frame, such that any pertinent detail that needs to be added to the representation can be added hierarchically under one of the existing nodes. The ontology is also used as a boundary object (i.e., an information object that facilitates discussion and interaction between divergent communities that share common interests but have different perspectives on those interests) so that the diverse stakeholders involved with the project can understand and agree on the features that must be taken into account when assessing system reliability and performance.

Building on the ontology, important features and relationships from the various existing representations (e.g., engineering diagrams, timeline and process diagrams) are integrated in a conceptual graph (or series of graphs). One of the strengths of conceptual graphs is that they are an effective common format to capture diverse concepts and relationships and thus provide an effective structure for combining information. If the concept or relationship can be described in natural language, it can be represented logically and eventually mathematically (the process works backward as well). The graphical statistical models developed in this process can be easily explained to stakeholder communities because they are representations of natural language in which relationships can be understood without having to explain the underlying mathematical and statistical notation.

Unlike reliability block diagrams and fault trees, conceptual graphs do not correspond directly to a particular statistical model. There must be a translation from the qualitative conceptual graph model to a quantitative model. Bayesian networks, in particular, are a flexible class of statistical graphical models that capture causal relationships (Jensen, 1996) in a way that meshes well with conceptual graphs; they are considered flexible because standard reliability diagrams (like block diagrams and fault trees) can easily be represented as Bayesian networks (Almond, 1995). The Bayesian network can also be used to model the conditional dependence and independence relationships important for specifying the more complex Bayesian, hierarchical, and random effects models mentioned earlier in this report. (For an example of the development of representations and the subsequent use of a Bayesian network for analysis, see Appendix C.)

Not every performance and reliability assessment requires the careful development of an integrated series of representations. However, these kinds of representations do help accomplish the goals of a system evaluation plan by making explicit the relationships among parts of the system and the analysis and by providing an explicit mapping of the evaluation to the data sources and the metrics of interest. The conceptual graph representations are flexible enough to achieve these goals within a framework that can change dynamically as the system and evaluation goals develop.

COMBINING INFORMATION FOR COMPLEX SYSTEMS

One of the steps required in an industrial reliability assurance program is the use of failure modes and effects analysis (FMEA) and reliability block diagrams to quantify the relationships between a system's subsystems, components, interfaces, and potential environmental effects. (As noted in the previous section, other representational methods can also be employed to create a unified picture of the system and decision space under consideration.) These representations result in a reliability model.

For large, complex, changing systems, however, developing and quantifying a reliability or performance model can be an extremely challenging problem. Complex systems tend to have complex problems, which usually exhibit one or more of the following characteristics (Booker and McNamara, 2003): a poorly defined or understood system or process, such as high cycle fatigue effects on a turbine engine; a process characterized by multiple exogenous factors whose impacts are not fully understood, such as the effects on a new system of changing combat missions; an engineered system in the very early stages of design, such as a new concept design for a fuel cell; a system, process, or problem that involves experts from different disciplinary backgrounds, who work in different geographical locations, and/or whose problem-solving tools vary widely (as is the case in the work involved to ensure the reliability of a manned mission to Mars); and any new groups of experts in novel configurations brought together for its solution.

Any time these sorts of complexities are involved, stakeholders may have difficulties coming to a common understanding of the problem to be addressed. As discussed previously, experts are always involved in the development and justification of assumptions used for modeling and analyses. In complex systems, one approach to dealing with the difficulties of formulating the model is to formalize the involvement of the experts. Briefly, the stages of expert involvement include the following:

1. Identifying the problem: What is the system under consideration? What are the primary metrics that must be evaluated? Who are the relevant stakeholders and what are their needs and expectations?

2. Operationalizing the problem: What are the operational definitions of the metrics? How will the metrics be evaluated? What are the constraints on the evaluation?

3. Developing the model: What are the core concepts that structure this problem? How are they related to one another? What classes of qualitative and quantitative models best fit the problem as currently structured?

4. Integrating and analyzing information: What data sources can be used to characterize this problem? Who owns those data sources? Can the postulated model answer the questions identified previously?

5. Statistical analysis: What are the appropriate techniques to combine the information from the available data sources? What are the appropriate graphical displays of information? What predictions or inferences must be made to support decisions?

NEED FOR ADDITIONAL STATISTICAL CAPABILITIES

The procedures, both informal and formal, for combining information comprise a broad set of techniques ranging from those that are extremely easy to apply (being robust to the precise circumstances of the application and with solutions in a simple, closed form) to very sophisticated models that are targeted to a specific application, require great imagination and technical expertise to identify and construct, and often require additional technical expertise to implement, possibly involving software development.[1] The more sophisticated category contains the rich collection of hierarchical and random effects models that have enjoyed recent, very rapid development and that have been successfully applied to a large number of new situations.

---

[1] Flexible, public-use software currently exists for a rich set of models, greatly simplifying software development. For example, R (http://cran.r-project.org/), BUGS (Bayesian Inference Using Gibbs Sampling), and WinBUGS (http://www.mrc-bsu.cam.ac.uk/bugs/winbugs/contents.shtml) are widely used in a variety of fields of application and are available at no cost on the Internet.

This report has described the potential applicability of various information-combining techniques to operational test design and evaluation for defense systems. Given the great complexity of ACAT I defense systems, and the many important facets of their development and evaluation, it is likely that many of these systems will require technically complicated methods to support their evaluation. It is also likely that use of these complicated methods will provide tangible benefits in operational evaluation. With respect to both hierarchical and random effects modeling, while there are some standard models that have been repeatedly applied and that may be useful for some defense applications, it is very likely that procedures at the leading edge of research will often be needed for high-quality operational test designs and evaluations.

Operational test evaluation is carried out under fairly substantial time pressures, in circumstances where errors can have extremely serious consequences. Experts with demonstrated proficiency in the use of combining information methods are required to ensure their fast, correct application. Furthermore, so that the ultimate decision makers can fully understand the findings based on these techniques, careful articulation of the methods and findings, including the important contribution of sensitivity analyses of divergence from assumptions, is very important, and also argues for the involvement of individuals with a complete understanding of the methods and their strengths and weaknesses.

Though there are exceptions, these sophisticated techniques are ordinarily not fully understood and correctly applied by those with a master's degree in statistics. A doctorate in statistics or a closely related discipline is generally required. This raises the question as to how ATEC can gain access to such expertise. The 1998 NRC report (National Research Council, 1998) mentioned available expertise at the Naval Postgraduate School, the Institute for Defense Analyses, RAND, and other federally funded research and development centers, as well as academia. This panel generally supports the recommendations contained in that report. One complication with the use of statisticians either on staff or, even more crucially, as consultants, is that more than statistical expertise will be required. Statisticians working on the methodology for system evaluation will need to work in close collaboration with experts in defense acquisition, military operations, and the system under test. Knowledge of physics and engineering would also be extremely useful.

Given the need for collaboration and acquired expertise, having appropriately qualified on-site staff is clearly the best option. Another option that should be considered, especially in the short term, is to offer sabbaticals and other temporary arrangements to experts in industry and academia. The panel suggests that one approach, which would institutionalize the use of sabbatical arrangements but that would require cooperation of the services, would be for each service to create an Interagency Personnel Act (PIA) position for a statistical expert for test and evaluation, reporting to the head of each service's operational test agency. DOT&E could also create a similar position. These statistical experts would work both as individual resident experts in each of the service test agencies, and would also be available to work jointly on the most challenging test and evaluation issues in the DoD. These temporary positions would rotate every three years and would have sufficient salary and prestige to attract leading statisticians from academia and industry.

In addition, ATEC should consider making available all sources and types of information for a candidate defense system to a selected group of qualified statisticians in industry and academia as a case study to understand the potential advantages of combining information for operational evaluation.

Recommendation: ATEC should examine how to increase statistical capabilities to support future use of techniques for combining information. As a first step, ATEC should consider providing all sources of information for a candidate defense system to a group of qualified statisticians in industry and academia as a case study to understand the potential advantages of combining information for operational evaluation.

5

Testing Challenges and Opportunities Posed by the Future Combat System

Although operational testing remains a challenge for the Stryker/SBCT, it will be distinctly easier than an operational test for the Army's Future Combat Systems (FCS), which is being developed under the Army's transformation program. This chapter provides background on the FCS-equipped objective force and highlights some challenges and opportunities for ATEC with respect to combining information for the FCS operational test and evaluation process.

All the U.S. military services are undergoing a transformation of their forces to provide a military capability to cover the spectrum of missions anticipated in the twenty-first century. The Army is pursuing the most ambitious of these transformations, as it changes from a predominantly heavy, forward-deployed force to one that is lighter and more strategically mobile for rapid and decisive operations anywhere in the world. It is advertised as a transformation from an industrial age force to a network-centric, information age force.

Stryker/SBCT is the interim part of this transformation program, which includes legacy, interim, and objective forces. Legacy forces are existing heavy forces (equipped with M-1 Abrams tanks and M-2 Bradley fighting vehicles) that are deployed overseas primarily by naval transport. Interim forces include a planned fielding of six SBCTs to air-deployable light divisions (e.g., the 82nd Airborne Division) by 2006 to provide them with more lethality and tactical mobility for lower-end operational missions. Objective forces will be FCS-equipped medium-weight forces developed to be rapidly deployable and capable of performing effectively over the full spectrum of anticipated operational missions in the twenty-first century.

Transformation to the objective force involves the development of new technologies and platforms, new operational concepts, and new force structures and organizations. The objective force under development is intended to replace all heavy legacy and interim forces in the Army. As currently planned, the first unit will be equipped in 2008, initial operational capability will occur in 2010, and three Future Brigade Combat Teams (FBCTs) will be fielded each year thereafter until all legacy and interim forces are replaced by 2032.

The linchpin of the objective force is the FCS, which is a family of systems involving many new technologies and platforms intended to make the force:

- fully deployable by C-130 aircraft by reducing vehicle weight to 16–20 tons;
- survivable, in spite of reduced armor protection, via a command, control, communications, computers, intelligence, surveillance, and reconnaissance (C4ISR) capability that provides information about the location of threat systems to a networked group of friendly systems; survivability will also be enhanced with active protection capabilities;
- lethal, through the use of precision munitions throughout the battlefield;
- sustainable with a reduced logistics footprint by having ultrareliable parts and systems; and
- versatile by providing capabilities necessary across the full spectrum of anticipated operational missions.

The FCS/FBCT will consist of up to two dozen new manned and unmanned systems. Manned ground systems include the infantry carrier vehicle, mounted combat system that can employ line-of-sight and beyond-line-of-sight fires, non-line-of-sight cannon system, non-line-of-sight mortar system, reconnaissance and surveillance vehicle, command and control vehicle, medical evacuation vehicle, and recovery vehicle.

Unmanned ground systems include the "missiles in a box" (Netfires) that remotely fire precision and loitering attack missiles, intelligent mine system that distributes mines on call, armed reconnaissance vehicle, small unmanned ground vehicle, and mule vehicle to help infantry carry materials. Unmanned aerial vehicles (UAVs) include the Class 1 lightweight UAV for close-in infantry use, Class 2 medium-range UAV for target acquisition and identification, Class 3 long-range UAV for target acquisition and identification, and Class 4 high-flier UAV for long-range use and command relay.

In addition, the C4ISR capability involves various unmanned sensors, new radios, new relays, and new architecture, protocols, and software for command-control, fire support, intelligence fusion, and management of the network. Because many of these systems depend on major advances in exploratory research and development programs and because of the Army's desire for initial operational capability in 2010, an evolutionary development process is being used. This includes spiral development for software and block improvements in component technologies every 3–5 years.

The Army is developing new force structures, organizations, and operational concepts for employment of the FCS-equipped objective force, all of which are still in their formative stages. The organizational basis for the FCS has been defined as a "unit of action." Although its composition is still being designed, the unit of action is viewed as comparable in size to today's brigade (and so the terms unit of action and FBCT are used synonymously here). The next echelon unit is referred to as the "unit of employment" and is viewed as comparable to today's division or corps. Concurrent with the development of the FCS, the Army is developing operational concepts for use of a net-centric FBCT, including new tactics, techniques, and procedures to exploit the new FCS technologies and obtain maximal effectiveness across the anticipated spectrum of operational missions.

TESTING CHALLENGES

The objective force will require extensive and continual testing and experimentation, using a process that is substantially more complex than the operational test conducted for the Stryker/SBCT.

The Stryker/SBCT was designed primarily for use in stability and security operations (SASOs) and, to a lesser extent, for the lower end of small-scale contingencies (SSCs), and the Stryker operational test program reflects this view. In contrast, the FCS/FBCT is being designed to perform over the full spectrum of operational missions and environments, and, accordingly, will likely require a commensurate breadth of operational testing. FCS/FBCT testing will likely involve performance and effectiveness measures relevant to major theater war, including force-on-force engagements against comparable and asymmetric threats; SSC operations (e.g., Panama); SASO activities including humanitarian assistance, peacekeeping, and postconflict infrastructure protection (as in Iraq); and counterterrorist conflicts for protection of U.S. military forces and installations and possibly homeland defense.

Although Stryker/SBCT was advertised as a system of systems, the operational test program does little to recognize or exploit this representation of Stryker. Given the extensive interdependencies among most of the manned systems and between the manned and unmanned ground and air systems in the FBCT, the Army will have to design and implement a feasible means of testing the effectiveness of the family of systems as well as the performance of individual systems in a broad range of operational missions and environments.

The potential success of the Army's transformation is more closely tied to the development of new operational concepts for employment of a net-centric force than to the development of new hardware. Experimentation and testing of these concepts and associated tactics, techniques, and procedures will be required as much for diagnostic purposes (e.g., what is not working and how can it be improved?) as for evaluation. Effective concepts and tactics, techniques, and procedures will likely differ with operational missions and environments and accordingly may need testing in a number of different settings. A critical component of the testing is to assess the degree to which the concepts make effective use of C4ISR assets and the tactical network and whether situation awareness will reduce the vulnerability of less-armored systems.

A major technical challenge for developers of the FCS/FBCT is the networking of all the system elements. The network must have the capacity and security to link all the ground and air systems; the manned and unmanned ones; those that conduct direct, beyond-line-of-sight, and non-line-of-sight fires; and numerous ground and air sensors, while many of these elements move around a constantly changing battlefield. It serves, in effect, as a mobile Internet. It is also an adaptive network that manages and allocates bandwidth where and when needed. Testing the effectiveness of such a network, its architecture, protocols, and the network management process in an operational environment (or multiple environments) will be equally challenging for the testing community.

As previously noted, initial operational capability for the FCS family of systems will be 2010 with continual significant block upgrades in capability every 3–5 years and changing software via a spiral development process. This will require a methodology for evolutionary test designs not previously explored in the testing community, such as, for example, a sequential, or perhaps continuous, testing of block upgrades that makes efficient use of previous test data to design subsequent tests and the effective combining of data to address subsequent test questions.

Reliability is a key performance parameter for the FCS. The Army claims that it is willing to sacrifice some system performance to obtain ultrareliability as a means to substantially reduce its logistic footprint. Operational testing for ultrareliability will be a significant challenge, requiring efficient combining of data and innovative approaches such as stress testing and accelerated testing.

TESTING OPPORTUNITIES

There is sufficient time before the first unit is equipped and initial operational capability is implemented for ATEC to begin planning the testing and evaluation program. This section lists some opportunities that might be useful in the planning process and eventual implementations, especially to address the challenges identified above.

As designed, the Stryker/SBCT operational test will assess the force's capabilities in a small number of average and not-too-stressful operational situations. An operational situation includes the operational mission, the relevant threat, and characteristics of the operational environment, elements of which are usually not under U.S. control. In contrast, the FCS/FBCT force will be employed in a broad spectrum of operational situations worldwide. Since resource constraints will limit testing and experimentation to a small number of situations, this limited testing should provide information that can be used to assess the force's capabilities over the broader range of situations. To accomplish this, the spectrum of operational situations should be described parametrically, creating a parametric operational situation (POS) space for use in the design, implementation, and analysis of the FCS/FBCT test and evaluation program. (Bonder, 1999, provides an example of the POS space concept.) The POS space, which comprises the set of likely operational situations to which the FCS/FBCT must have the capability to respond successfully, will enable the Army to identify and select test points that are stressful so that assessments of capabilities for the other untested points in the space are interpolations rather than extrapolations of the tested points. Use of the POS space will also enable the Army to both assess the degree of versatility (an advertised capability) that the

FCS/FBCT will bring to situations in the space and presumably identify ways to make it more versatile. Creating the POS space will require the development of a broad taxonomy for the space as well as dimensions for each category of the taxonomy, metrics for the dimensions, and possible value ranges for each of the metrics. Real-world realizations in the space can then be represented by values for the dimension metrics of each taxonomy category.

Testing and related activities performed over many years will provide a stream of data that should be used to periodically diagnose and evaluate the FCS/FBCT as it is fielded in an evolutionary development and procurement process. To make effective and efficient use of the data, ATEC should design and implement an FCS/FBCT archive for data generated over time in tests, experiments, field use, and related analysis and evaluation activities. Such an archive will provide a centralized location of FCS/FBCT performance and effectiveness data for different operational situations generated over time by developmental tests, operational tests and experiments, field use, and simulation-based analyses. If it is structured to make effective use of the latest data-mining techniques, the archive should:

- facilitate effective combining of data from different sources over time;
- provide information needed to interpolate the results of testing in a limited number of operational situations in order to assess capabilities of the FCS/FBCT in situations not tested;
- provide a "hot plant" for reliability assessments;
- provide data to design efficient subsequent testing in the FCS/FBCT evolutionary development process;
- assist in developing performance correlates (e.g., reliability) for design of FCS evolutionary upgrades;
- provide data to assess the versatility of FCS/FBCT worldwide; and
- provide data to compare the results of simulation analyses with those of operational testing.

The Army is still in the process of developing models and simulations to support development and analyses of the FCS program. ATEC should coordinate its efforts with those of the modeling and simulation program to ensure that the models can be used to support the testing and evaluation program. Specifically, the models and simulations should be the basis of analyses that will be used to help select appropriate operational situations for testing; determine appropriate issues and hypotheses for testing; guide test designs; assist in analyses and extensions of test data; and provide a broader assessment of FCS capabilities, given test data from a limited number of operational situations.

STRATEGY FOR TESTING AND EVALUATION

ATEC should prepare a strategy (not just a test design) to overcome the testing and evaluation challenges noted above, and this strategy should recognize the sequential nature of the testing that will be required as part of the evolutionary acquisition process for FCS. There is currently a strong emphasis in defense acquisition on acquiring new systems in stages, an approach known as evolutionary or block acquisition, with each stage undergoing separate development, testing, and evaluation. The staging reduces the ultimate risk of acquiring a deficient system, and the maturation of the system at each stage reduces the problems often faced by integration of newly developed components. This change in emphasis has a number of implications, especially with respect to experimental designs that are specifically suited to staged development. In addition, there are implications with respect to both the utility of statistical approaches to combine information for system evaluation at each stage, and the increased need for data archives to support this type of system evaluation and the experimental design used for each subsequent stage.

The strategy should also recognize both the need to evaluate the family of systems and the potential need for diagnostic experimentation of operational concepts in multiple operational situations. It should delineate relevant questions to be addressed by testing and evaluation (perhaps as strategy objectives) and by the combining of information, not only for evaluations but also to determine the value of additional data from subsequent tests to address these questions. The strategy should include modeling and simulation activities as an integral part of the test design and evaluation process, mindful of the need for validation.

Recommendation: ATEC should prepare a strategy for operational testing of the FCS/FBCT that will:
- **recognize the sequential nature of the testing that will be required as part of the evolutionary acquisition process for FCS,**

- recognize the need to evaluate the family of systems and the potential need for diagnostic experimentation of operational concepts in multiple operational situations,
- delineate relevant questions to be addressed by testing and evaluation,
- determine the value of additional data (from subsequent tests) needed to address these questions, and
- include modeling and simulation activities as an integral part of the testing and evaluation process.

References

Abernethy, R.B., J.E. Breneman, C.H. Medlin, and G.L. Reinman
    1983   *Weibull Analysis Handbook*. Air Force Wright Aeronautical Laboratories Technical Report AFWAL-TR-83-2079. Available from the National Technical Information Service, Washington, DC.

Almond, R.
    1995   *Graphical Belief Modeling*. London: Chapman and Hall.

Arcones, M.A., P.H. Kvam, and F.J. Samaniego
    2002   Nonparametric estimation of a distribution subject to a stochastic precedence constraint. *Journal of the American Statistical Association* 97:170–182.

Ascher, Harold, and Harry Feingold
    1984   *Repairable Systems Reliability: Modeling, Inference, Misconceptions, and Their Causes*. Newbury, UK: Marcel Dekker.

Belsley, David A., Edwin Kuh, and Roy E. Welsch
    1980   *Regression Diagnostics: Identifying Influential Observations and Sources of Collinearity*. New York: John Wiley and Sons.

Berger, J.O., and D.A. Berry
    1988   Statistical analysis and the illusion of objectivity. *American Scientist* 76:159–165.

Bonder, Seth
    1999   Versatility Planning: An Idea Whose Time Has Come ... Again! Steinhardt Lecture presented at the Second International Military Applications Society Conference, El Paso.

Booker, J.M., and L.A. McNamara
    2003   Expertise and Expert Judgment in Reliability Characterization: A Rigorous Approach to Eliciting, Documenting, and Analyzing Expert Knowledge. *Engineering Design Reliability Handbook*. D. Ghiocel and E. Nikolaides, eds. Boca Raton, FL: CRC Press.

Bordley, R.F.
- 1998 R&D project generation versus R&D project selection. *IEEE Transactions in Engineering Management* 45:407–413.

Box, G.E.P., and J.S. Hunter
- 1961 The 2^{k-p} fractional factorial designs. *Technometrics* 3:311–351.

Box, G.E.P., W.G. Hunter, and J.S. Hunter
- 1978 *Statistics for Experimenters*. New York: John Wiley and Sons.

Carlin, B.P., and T.A. Louis
- 1996 *Bayes and Empirical Bayes Methods for Data Analysis*. London: Chapman and Hall.

Chaloner, Kathryn
- 1984 Optimal Bayesian experimental design for linear models. *Annals of Statistics* 12:283–300.

Coleman, D.E., and D.C. Montgomery
- 1993 A systematic approach to planning for a designed experiment. *Technometrics* 35(1):1–12.

Gaver, Donald P., Patricia A. Jacobs, and Arthur Fries
- 1997 Prediction of Changeover Performance: Operational Test (OT) Rates from Developmental Test (DT) Rates via Meta-Analysis. *Proceedings of the Section on Government Statistics and Section on Social Statistics of the American Statistical Association*.

Gelman, Andrew, John B. Carlin, Hal S. Stern, and Donald B. Rubin
- 1995 *Bayesian Data Analysis*. New York: Chapman and Hall.

Hahn, G.J.
- 1993 Discussion: Experimental design review and comment. *Technometrics* 35:15–17.

Hollander, M., and D.A. Wolfe
- 1973 *Nonparametric Statistical Methods*. New York: John Wiley and Sons.

Jensen, F.
- 1996 *An Introduction to Bayesian Networks*. New York: Springer-Verlag.

Kahneman D., P. Slovic, and A. Tversky (eds.)
- 1982 *Judgment and Uncertainty: Heuristics and Biases*. Cambridge, UK: Cambridge University Press.

Kiefer, Jack
- 1959 K-Sample Analogues of the Kolmogorov-Smirnov and Cramer-v. Mises Tests. *Annals of Mathematical Statistics* 30:420–446.

Leishman, D., and L. McNamara
- 2002 Interlopers, Translators, Scribes, and Seers: Anthropology, Knowledge Representation and Bayesian Statistics for Predictive Modeling in Multidisciplinary Science and Engineering Projects. Presented at the Conference on Visual Representations and Interpretations 2002, Liverpool, UK.

McLean, H.
- 2000 *HALT, HASS and HASA Explained: Accelerated Reliability Techniques*. Milwaukee, WI: ASQ Press.

Meeker, W.Q., and L.A. Escobar
- 1998 *Statistical Methods for Reliability Data*. New York: John Wiley and Sons.

Meyer, M.A., and J.M. Booker
  2001 *Eliciting and Analyzing Expert Judgment: A Practical Guide.* ASA-SIAM Series on Statistics and Applied Probability, Volume 7.

Meyer, M.A., and R.C. Paton
  2002 Interpreting, Representing and Integrating Scientific Knowledge from Interdisciplinary Projects. *Theoria et Historia Scientiarum* 6(2):323–356.

National Research Council
  1992 *Combining Information: Statistical Issues and Opportunities for Research.* Committee on Applied and Theoretical Statistics. D. Gaver et al. eds. Washington, DC: National Academy Press.
  1998 *Statistics, Testing, and Defense Acquisition: New Approaches and Methodological Improvement.* Panel on Statistical Methods for Testing and Evaluating Defense Systems, Committee on National Statistics, Commission on Behavioral and Social Sciences and Education. Michael Cohen, John Rolph, and Duane Steffey, eds. Washington, DC: National Academy Press.

Nelson, Wayne
  1985 Weibull analysis of reliability data with few or no failures. *Journal of Quality Technology* 17:140–146.
  2003 *Recurrent Events Data Analysis for Product Repairs, Disease Recurrences, and Other Applications.* ASA-SIAM, Philadelphia.

Paton, R.C., S. Lynch, D. Jones, H.S. Nwana, T.J.M. Bench-Capon, and M.J.R. Shave
  1994 Domain Characterisation for Knowledge Based Systems. *Proceedings of A.I. 94—Fourteenth International Avignon Conference*, Volume 1, 41–54.

Proschan, F.
  1963 Theoretical explanation of observed decreasing failure rate. *Technometrics* 5:375–383.

Reese, C.S., A.G. Wilson, M.S. Hamada, H.F. Martz, and K. Ryan
  2000 Integrated Analysis of Computer and Physical Experiments. Los Alamos Unclassified Report LA-UR-00-2915. To appear *Technometrics.*

Samaniego, F.J., D. Steffey, and H. Tran
  2001 *Towards a Theory for Combining Information from "Related" Experiments.* Technical Report. Department of Statistics, University of California, Davis.

Samaniego, F.J., and E. Vestrup
  1999 On improving upon standard estimates via linear empirical Bayes methods. *Statistics and Probability Letters* 44:309–318.

Scholz, F.W., and M.A. Stephens
  1987 K-Sample Anderson-Darling Tests. *Journal of the American Statistical Association* 82:918–924.

Smith, A.F.M., and A.E. Gelfand
  1992 Bayesian statistics without tears: A sampling-resampling perspective. *The American Statistician* 46:84–88.

Sowa, J.
  1984 *Conceptual Structures.* Reading, MA: Addison-Wesley.

Tversky, A., P. Slovic, and D. Kahneman, (eds)
  1985 *Judgment Under Uncertainty: Heuristics and Biases.* Cambridge, MA: Cambridge University Press.

United States Department of Defense
- 1993 MIL-STD-690C: "Failure Rate Sampling Plans and Procedures," U.S. Department of Defense, Washington, DC.
- 1977 MIL-STD-781C: "Reliability Design Qualifications and Production Acceptance Tests: Exponential Distributions," U.S. Department of Defense, Washington, DC.
- 1960 MIL-HDBK-108: "Quality Control and Reliability—Sampling Procedures and Tables for Life and Reliability Testing (Based on Exponential Distribution)," U.S. Department of Defense, Washington, DC.

Wu, C.F.J., and M.S. Hamada
- 2000 *Experiments: Planning, Analysis and Parameter Design Optimization.* New York: John Wiley and Sons.

Appendix A

Further Details Concerning the Bearing Cage Example

In Chapter 3, an overview of a frequentist analysis of the reliability of a bearing cage was provided, starting with the analysis when the two Weibull parameters, η and β, are both estimated, and then demonstrating the benefits from using other sources of information to fix β. This was followed by a Bayesian analysis in which other sources of information were used to provide a prior for β.

With respect to the frequentist analysis, this appendix provides additional probability plots for the bearing cage data, fixing β at the values 1.5, 2, and 3. These plots show the lack of support for the assertion that B10 is 8,000 hours.

With respect to the Bayesian analysis, this appendix provides two plots, the first of the prior distribution for B10 and β, and the second for the posterior distribution for B10 and β. Both plots have the data likelihood superimposed. It is clear that the data likelihood has moved the posterior to be more consistent with it than the prior.

Figures A-1, A-2, and A-3 show probability plots with the Weibull shape parameter β fixed at 1.5, 2, and 3, respectively. Although Figure A-1 may allow for a little optimism, the overall conclusion suggested by these figures is that the bearing cage is, most likely, not meeting its reliability goal.

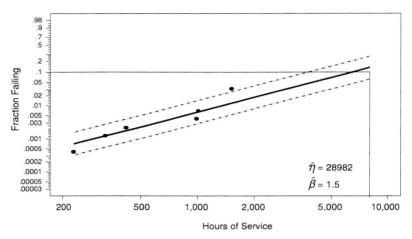

FIGURE A-1 Weibull probability plot of the bearing cage failure data showing the maximum likelihood estimate of fraction failing with fixed $\beta = 1.5$, the reliability target, and approximate confidence limits. SOURCE: Meeker and Escobar (1998).

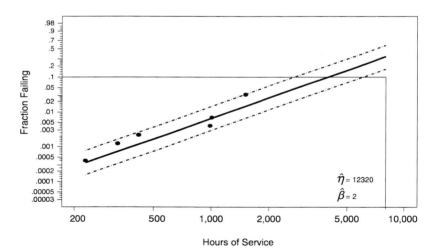

FIGURE A-2 Weibull probability plot of the bearing cage failure data showing the maximum likelihood estimate of fraction failing with fixed $\beta = 2$, the reliability target, and approximate confidence limits. SOURCE: Meeker and Escobar (1998).

THE BEARING CAGE EXAMPLE

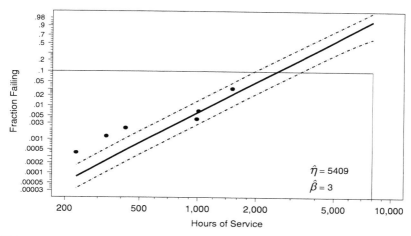

FIGURE A-3 Weibull probability plot of the bearing cage failure data showing the maximum likelihood estimate of the fraction failing with fixed $\beta = 3$, the reliability target, and approximate confidence limits. SOURCE: Meeker and Escobar (1998).

BAYES ANALYSIS

Figure A-4 shows a sample of points from the joint prior distribution of β and B10 superimposed on a graph of the relative likelihood contours. The truth is expected to lie at the intersection of the prior and the likelihood.

A sample from the posterior can be obtained by filtering the sample from the prior distribution. This can be done by selecting points in the prior distribution with a probability equal to the likelihood contour going through the point. Points are run through this filter until a sufficient number of posterior points are obtained to estimate the posterior distribution (for two parameters, something on the order of 6,000 to 10,000 is sufficient).

Figure A-5 shows a sample of points from the joint posterior distribution, again with the likelihood contours.

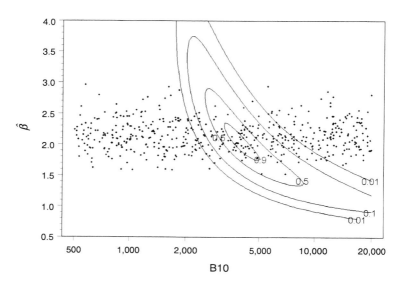

FIGURE A-4 Sample points from the joint *prior* distribution of B10 of bearing cage life and the Weibull β with the data likelihood superimposed. SOURCE: Meeker and Escobar (1998).

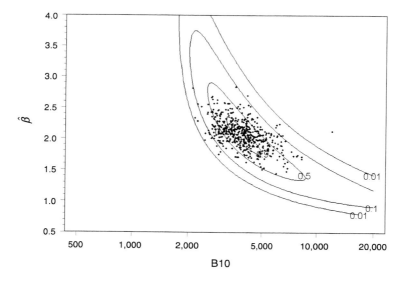

FIGURE A-5 Sample points from the joint *posterior* distribution of B10 of bearing cage life and the Weibull β with the data likelihood superimposed. SOURCE: Meeker and Escobar (1998).

Appendix B

Technical Details on Combining Information in Estimation: A Treatment of Separate Failure Modes

The last section of Chapter 2 indicated an approach to a part of the following general problem: How can one use information from developmental testing to reveal the reliability of a defense system in operational testing when the failure modes of developmental and operational testing may be distinct? The discussion in Chapter 2 outlined the methods for a simple special case, leaving the technical details for this appendix.

Consider the following problem. In developmental testing, a system exhibits two failure modes of interest, and the distance to failure (in miles) for each mode is an exponentially distributed random variate, with (unknown) failure rates Λ_1 and Λ_2, respectively. In operational testing, we assume that one of four possibilities remains: (1) all failure modes have been removed, (2, 3) either of the individual failure modes remains, or (4) both failure modes remain. We would like to use operational test data to update expert judgment concerning the probability of each of these four possible situations.

PRIOR INFORMATION

Prior to the start of operational testing, we assume that previous engineering experience, analysis of developmental test data, and redesign actions support the assessment of an a priori prior probability density function (p.d.f.) for each failure rate:

- $f_{\Lambda_1}(\lambda_1)$
- $f_{\Lambda_2}(\lambda_2)$

In other words, the values provided by these functions are expert assessments of the probable values of the two failure rates.

In addition, the same tests and judgments support the assignment of a priori probabilities for four possible scenarios:

- p_0 = probability {no failure modes remain}
- p_1 = probability {only mode 1 remains}
- p_2 = probability {only mode 2 remains}
- $p_3 = 1 - p_0 - p_1 - p_2$ = Probability {both mode 1 and mode 2 remain}

For the purpose of this exposition, we assume that Λ_1 and Λ_2 are independent, and their priors can each be well represented by a Gamma p.d.f.: $f_{\Lambda_1}(\lambda_1) = g(\lambda_1 \mid a_1, b_1)$ and $f_{\Lambda_2}(\lambda_2) = g(\lambda_2 \mid a_2, b_2)$, where $g(x \mid a, b) = \dfrac{b^a x^{a-1} e^{-bx}}{\Gamma(a)}$. In the analysis below, the following properties of the Gamma p.d.f. are applied:

Moments of the Gamma p.d.f.:

$$E(X) = \frac{a}{b}; \quad Var(X) = \frac{a}{b^2}; \quad E(X^2) = \frac{a(a+1)}{b^2};$$

$$E\left(\frac{1}{X}\right) = \frac{b}{a-1}; \quad Var\left(\frac{1}{X}\right) = \frac{b^2}{(a-1)^2(a-2)}; \quad E\left[\left(\frac{1}{X}\right)^2\right] = \frac{b^2}{(a-1)(a-2)};$$

$$E(e^{-Xt}) = \left(\frac{b}{b+t}\right)^a; \quad Var(e^{-Xt}) = \left(\frac{b}{b+2t}\right)^a - \left(\frac{b}{b+t}\right)^{2a};$$

$$E\left[\left(e^{-Xt}\right)^2\right] = \left(\frac{b}{b+t}\right)^a$$

POSTERIOR COMPUTATION

Assume the system has a failure mode with distance to failure being exponentially distributed with parameter Λ (so that the number of failures in a test interval t is Poisson distributed with parameter Λt), and assume that Λ has prior p.d.f. $f_\Lambda(\lambda) = g(\lambda | a', b')$.

Assuming that testing yields n failures in t miles, then the posterior p.d.f. of Λ is $g(\lambda | a'', b'')$, where the "updated" parameters are $a'' = a' + n$, $b'' = b' + t$.

UNCONDITIONAL DISTRIBUTION OF THE NUMBER OF FAILURES IN A TEST INTERVAL

If a failure rate is Λ, where p.d.f. $g(\lambda | a, b)$, then $n =$ the number of failures (for a single, unidentified mode) during an exposure of t (miles) and has the following unconditional probability mass function:

$$f_N(n) = \int_0^\infty \frac{(\lambda t)^n}{n!} e^{-\lambda t} g(\lambda | a, b) \, d\lambda = h\left(n \middle| a, \frac{t}{t+b}\right)$$

where

$$h\left(n \middle| a, \frac{t}{t+b}\right) \equiv \binom{n+a-1}{a-1}\left(\frac{b}{t+b}\right)^a \left(\frac{t}{t+b}\right)^n \quad n = 0, 1, 2 \text{ (the negative binomial distribution).}$$

AN APPROXIMATION FOR THE DISTRIBUTION OF THE SUM OF TWO GAMMA DISTRIBUTED VARIATES

Although there is no closed-form expression for the p.d.f. of the sum of two gamma distributed random variables, a reasonable approximation can be obtained. In particular, if $f_{X_1}(x) = g(x | a_1, b_1)$, $f_{X_2}(x) = g(x | a_2, b_2)$, and $X_3 = X_1 + X_2$, then $f_{X_3} \approx g(x | a_3, b_3)$, where

$$b_3 = \frac{\frac{a_1}{b_1} + \frac{a_2}{b_2}}{\frac{a_1}{b_1^2} + \frac{a_2}{b_2^2}}, \text{ and } a_3 = b_3\left(\frac{a_1}{b_1} + \frac{a_2}{b_2}\right)$$

PRIOR PERFORMANCE MEASURES

Without data from operational testing, the probability assessments above enable test evaluators to evaluate typical performance measures of interest. In particular, using the information for the moments of the Gamma distribution given above, and the approximation above for the sum of two independent Gamma random variables, these can be shown to be:

a) $\hat{\lambda}$, an estimate of the system's total failure rate, Λ, and its uncertainty (expressed by its standard deviation, $\sigma_\Lambda = \sqrt{\text{Var}(\Lambda)}$),

$$\hat{\lambda} = \text{E}(\Lambda) = \sum_{i=1}^{3} p_i \left(\frac{a_i}{b_i}\right)$$

$$\sigma_\Lambda^2 = \sum_{i=1}^{3} p_i \left(\frac{a_i(1+a_i)}{b_i^2}\right) - \hat{\lambda}^2$$

b) $\hat{\mu}$, an estimate of the system's mean miles to first failure, and its uncertainty σ_μ

$$\hat{\mu} = \sum_{i=1}^{3} p_i \left(\frac{b_i}{a_i - 1}\right)$$

$$\sigma_\mu^2 = \sum_{i=1}^{3} p_i \left(\frac{b_i^2}{(a_i - 1)(a_i - 2)}\right) - \hat{\mu}^2$$

c) $\hat{r}$, the expected value of the reliability of the system at t miles (e.g., the probability that there is no failure in the first t miles), and its uncertainty σ_R

$$\hat{r} = E[R(M)] = \sum_{i=1}^{3} p_i \left(\frac{b_i}{b_i + M}\right)^{a_i}$$

$$\sigma_r^2 = \sum_{i=1}^{3} p_i \left(\frac{b_i}{b_i + 2M}\right)^{a_i} - \hat{r}^2$$

COMBINING PRIOR INFORMATION WITH DATA FROM OPERATIONAL TESTING

After t test miles in operational testing, if the data show n_1 mode 1 failures and n_2 mode 2 failures, then the performance measures can be calculated as above, but with the parameters "updated" so that a_i and b_i are replaced by a'_i and b'_i, respectively, where:

$$a'_i = a_i + n_i; \quad b'_i = b_i + t, \quad i = 1,2$$

and the posterior probabilities for the failure mode scenarios are:

$$p'_0 = \begin{cases} \dfrac{p_0}{A(0,0)} & n_1 = n_2 = 0 \\ 0 & \text{else} \end{cases}$$

$$p'_1 = \begin{cases} \dfrac{1}{A(n_1,0)} p_1 h\left(n_1 \middle| a'_1, \dfrac{t}{t+b'_1}\right) & n_1 > 0, n_2 = 0, \\ 0 & \text{else} \end{cases}$$

$$p'_2 = \begin{cases} \dfrac{1}{A(0,n_2)} p_2 h\left(n_2 \middle| a'_2, \dfrac{t}{t+b'_1}\right) & n_1 = 0, n_2 > 0 \\ 0 & \text{else} \end{cases}$$

$$p'_3 = \begin{cases} \dfrac{1}{A(n_1,n_2)} p_3 h\left(n_1 \middle| a'_1, \dfrac{t}{t+b'_1}\right) h\left(n_2 \middle| a'_2, \dfrac{t}{t+b'_2}\right) & n_1 > 0, \; n_2 > 0 \\ 0 & \text{else} \end{cases}$$

where

$$A(0,0) = p_0 + p_1 h\left(0 \middle| a'_1, \dfrac{t}{t+b'_1}\right) + p_2 h\left(0 \middle| a'_2, \dfrac{t}{t+b'_2}\right) + p_3 h\left(0 \middle| a'_1, \dfrac{t}{t+b'_1}\right) h\left(0 \middle| a'_2, \dfrac{t}{t+b'_2}\right)$$

$$A(n_1,0) = p_1 h\left(n_1 \middle| a'_1, \dfrac{t}{t+b'_1}\right) + p_3 h\left(n_1 \middle| a'_1, \dfrac{t}{t+b'_1}\right) h\left(0 \middle| a'_2, \dfrac{t}{t+b'_2}\right) \quad n_1 > 0$$

$$A(0,n_2) = p_2 h\left(n_2 \middle| a'_2, \dfrac{t}{t+b'_2}\right) + p_3 h\left(0 \middle| a'_1, \dfrac{t}{t+b'_1}\right) h\left(n_2 \middle| a'_2, \dfrac{t}{t+b'}\right) \quad n_2 > 0$$

$$A(n_1,n_2) = p_3 h\left(n_1 \middle| a'_1, \dfrac{t}{t+b'_1}\right) h\left(n_2 \middle| a'_2, \dfrac{t}{t+b'_2}\right) \quad n_1 > 0, \; n_2 > 0$$

Appendix C

The Rocket Development Program

As an illustration of the methods Los Alamos National Laboratory is developing to evaluate the nuclear stockpile, consider the development program for a ballistic missile target for air defense system testing, referred to as the Rocket Development Program (RDP). The oversight agency for the RDP is the Rocket Development Program Center (RDPC), which is primarily responsible for project management, cost controls, and scheduling. Two groups of engineers are responsible for building separate sections of the rocket: one group is building a booster to send the rocket into the upper atmosphere, and the other is designing test payload for the rocket. Several subcontractors and vendors provide parts and support to each of the two primary engineering groups.

The RDPC program managers must predict performance and reliability for a system that is still in the design stages, determine whether the system will operate effectively when flown, and identify early any areas of technical risk. These efforts are complicated by the following facts: the rocket development program is extremely expensive; only one or two are built and flown, and they are usually destroyed in the process; and the engineers are rarely able to salvage subsystems for reuse in further iterations of the program.

Because each system flown is unique, there are few direct performance or reliability data available for parts or subsystems on the test rocket. Therefore, the important goals for the program are to collect data to help the air defense systems understand their likely performance against targets, and to fly a trajectory that falls within certain parameters. Accomplishment of both these goals constitutes mission success.

Visual representations for the RDP were developed using the conceptual graph techniques of Sowa (1984), whose approach combines a mapping to and from natural language with a mapping to logic. A conceptual graph, which consists of concepts and relations connected by arcs, illustrates a proposition with a finite connected bipartite graph, where concepts represent any entity, attribute, action, state, or event that can be described in natural language; relations detail the roles that each concept plays; and the arcs serve as connectors between the two. Figure C-1 shows a top-level ontology developed with a conceptual graph representation. In this example, concepts are shown as rectangles, relations as circles, and the relationships between the two as arcs. The ontology captures the basic cognitive categories of information about the RDP.

Identifying such categories makes it possible to ask questions about a system, even when one is not an expert. In the RDP example illustrated in Figure C-1, the ontology reveals key focus areas, such as: What functions were required in order for a particular mission event to occur? What parts were required for the function to occur? The ontology also differentiates between two stages in the design process: design time, when the engineers are working to plan and build the rocket; and run time, which represents the actual functioning of the rocket during flight.

The ontology developed in Figure C-1 is much too high level to directly support quantitative model development. Instead, it guides the elicitation of expertise necessary to gather the information required for developing quantitative models and metrics. After the ontology is developed, one can begin to develop specific representations for each of the concepts—for example, the parts and functions required to instantiate an event.

Once a preliminary representation of the important concepts has been developed, one of the most difficult tasks is operationalizing the evaluation metrics. In order to operationalize metrics such as *collecting sufficient data, flying a correct trajectory,* and *mission success* for the RDP, the analyst meets with the project leaders to identify specific goals for the rocket system, to describe an overview of how the rocket would function, and to find out which contractors were responsible for the major areas of the project. For example, flying a correct trajectory involves reaching apogee between 150 and 160 seconds after launch; and collecting sufficient data requires the forward cameras to operate with less than 10 seconds of data loss.

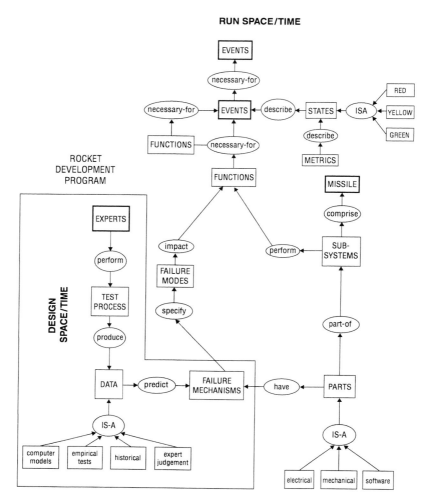

FIGURE C-1   Ontology for RDP.
SOURCE: Leishman and McNamara (2002).

During problem definition, a great deal of information is collected from a variety of stakeholders in the program. A number of tools can be used to structure this information. The first goal during model development for a large, complex system is to develop a qualitative map of the problem space that can be used to develop appropriate quantitative models. This map is often called a knowledge model and can be thought of as an elaboration of parts of the ontology. Graphical representations are used because most people find them easy to understand and because they are used commonly in many communities (e.g., engineering drawings). In large and complex problems, there are often many communities working on the problem—engineers, physicists, chemists, materials scientists, statisticians, tacticians, field commanders—and each has its own view of the problem definition and solution. The goal of the development of qualitative maps is to come to a common set of representations that allows everyone to have a shared understanding of the problem to be solved. Two expansions of the initial ontology are given in Figures C-2 and C-3.

In the RDP example, the first specific representation discretized the flight-time events required to fly a threat-representative (TR) trajectory (Figure C-2). Once these events were identified explicitly, they could be mapped into their importance for mission success. Subsequently, each event was represented by three diagrams at a finer level: a functional diagram (Figure C-3) that detailed only the functions required for an event; a subsystem-part diagram that broke subsystems into collections of parts; and a modified series parallel diagram that specified the order in which parts of a subsystem work together to perform a function. Figure C-3 identifies two primary functions for TR flight, data collection/vehicle tracking and boosted flight, which are themselves broken into several subfunctions. These subfunctions, in turn, can be further specified by the parts and subsystems involved in their performance. The diagrams are important because they help identify the dependencies that will have to be represented in the statistical model.

Definition of the levels that will be included in the problem must be related to the goals. For example, a decision maker may need only a rough comparison of a new design to the old in order to answer the question, "Is the new one at least as good as the old one?" In this case, it may not be necessary to represent the structure of the two systems down to the parts. The extent of information availability, including data and experts, can dictate how levels of detail are identified and chosen for inclusion in the model. For example, if knowledge is completely lacking at a subcomponent level, the problem should be defined and structured at the component level.

Once sufficient granularity has been achieved in the qualitative maps of the problem, the translation to quantitative models is possible. Since the qualitative maps are graphical, it is often helpful to develop graphical representations for the quantitative models as well—for example, reliability block diagrams, fault trees, and Bayesian networks. The Bayesian network shown

94

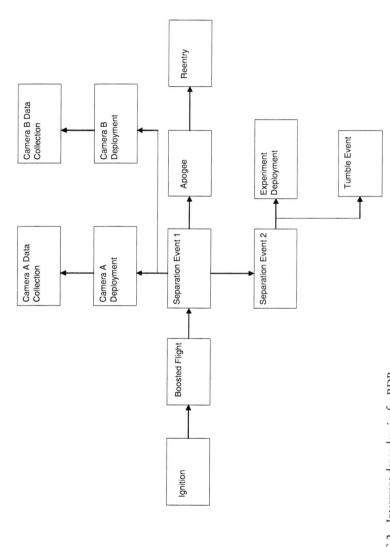

FIGURE C-2 Interevent dependencies for RDP.
SOURCE: Leishman and McNamara (2002).

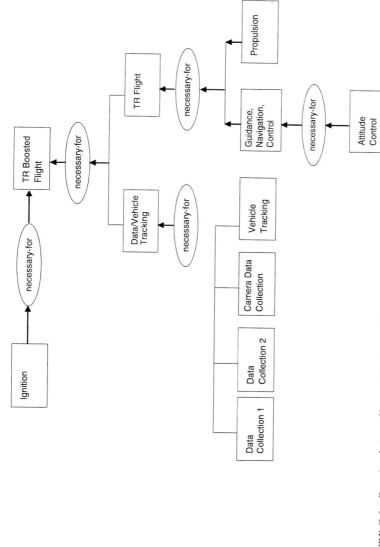

FIGURE C-3 Functional view of boosted flight for RDP.
SOURCE: Leishman and McNamara (2002).

in Figure C-4 captures a small part of the quantitative structure for the information about events, functions, and parts needed to quantify the model; Figure C-5 is a more traditional Bayesian network that consists of data and parameters.

While the quantitative model is being developed, it is important to examine potential data sources. What data are available to populate the model? Who owns the data? Perhaps most importantly, can the data and the model be used to answer the questions and evaluate the metrics from the problem identification stage of the analysis? One of the features of large and complex problems is the heterogeneity of data sources. Seldom will there be enough designed experimental data to evaluate each metric; consequently, additional sources of information must be used, such as computer models, engineering judgment, historical data, and developmental test data. Table C-1 is a sample of the kinds of data available to populate the Bayesian networks shown in Figures C-4 and C-5.

The heterogeneity of the data requires statistical methodological development to integrate the data and achieve appropriate estimates of uncertainty. The extensive modeling described in previous sections of this report makes explicit where and how the diverse data sources are being used in

TABLE C-1 Data Available for RDP Bayesian Network

Engineering Judgment
- The probability of the motor mount ring failing catastrophically is under 1 percent.
- If the motor mount ring fails catastrophically, then the fins and frame fall off the vehicle.
- There is somewhere between a 5 percent and 10 percent chance that the skin will peel back.
- If the fins or frame are missing, then the vehicle is unstable.
- If the skin peels back, then the vehicle is unstable.
- If the fins warp, then vehicle stability is compromised.

Experimental Data
- There is about a 10 percent chance that the fins will warp during flight.
- The frame will not fail if loads do not exceed 5,000 psi.

Computer Model
- Simulations indicate that there is a 15 percent chance that flight loads exceed 5,000 psi.

ROCKET DEVELOPMENT PROGRAM

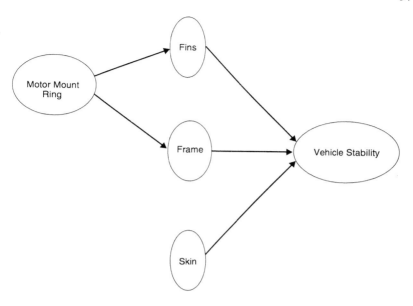

FIGURE C-4   Bayesian network.

support of the analysis. The questions for the statistical analysis are standard: What are the appropriate techniques to combine the available data sources? What are the appropriate graphical displays of information? What predictions or inferences must be made to support decisions?

For the RDP problem, the Bayesian network had over 2,000 nodes. Once the data had been identified and mapped to the structure, Markov chain Monte Carlo techniques were used to make a variety of estimates. In particular, RDPC was interested in the probability of mission red (complete failure), yellow (partial failure), or green (success). Initially, these were estimated to be 15 percent (± 5 percent), 60 percent (± 10 percent), and 25 percent (± 5 percent), respectively. However, estimates were available (with associated uncertainties) for probabilities of success and failure of components and functions throughout the system, and these were used to determine where further testing could be of value in increasing the probability of mission green and in decreasing the uncertainty of the estimates.

Not every performance and reliability assessment requires the careful development of a knowledge model. However, for large, complex systems

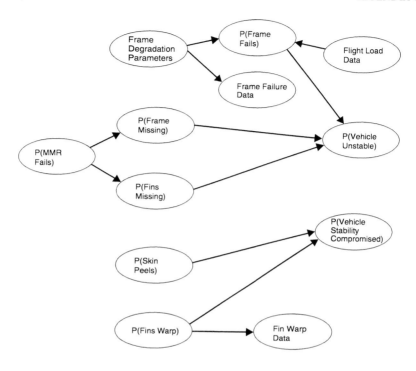

FIGURE C-5 Bayesian network with statistical parameters and data.

with heterogeneous data sources, the development of a common set of representations has many advantages: the representations provide a common language for all communities to interact with the problem, they can be used to explicitly identify the heterogeneous data sources, and they show an explicit mapping from the problem to the data to the metrics of interest.

Appendix D

Acronyms and Abbreviations

ABCS	Army Battle Command Systems
AoA	add-on armor
ATEC	Army Test and Evaluation Command
ATIRS	army test incident reporting system
BRL/AMSAA	Ballistics Research Laboratories/U.S. Army Materiel Systems Analysis Activity
C4ISR	command, control, communications, computers, intelligence, surveillance, and reconnaissance
CAA	Center for Army Analysis
CCTT	close combat tactical training
CF	casualty frequency
DoD	Department of Defense
DOE	Department of Energy
DT	Developmental test
DTP	detailed test plan
DOT&E	director, operational test and evaluation
FBCB2	Force XXI Battle Command Brigade and Below
FBCT	Future Brigade Combat Team
FCS	Future Combat System

FDSC	failure definition and scoring criteria
FLIR	forward-looking infrared systems
GFE	government-furnished equipment
IAV	interim armored vehicle
IBCT	Interim Brigade Combat Team
ICV	infantry carrier vehicle
IOT	initial operational test
LANL	Los Alamos National Laboratory
LER	loss exchange ratio
LIB	Light Infantry Brigade
MEP	Mission Equipment Package
MIB	Mechanized Infantry Brigade
MIL-STD	military standard
MOE	measure of effectiveness
MOP	measure of performance
MR	maintenance ratio
MTF	mean time to failure
MTTR	mean time to repair
NBC	nuclear/biological/chemical
NRC	National Research Council
OMS/MP	operational mission summary/mission profile
OPFOR	opposition force
ORD	Operational Requirements Document
OT	operational tests
PMCS	preventive maintenance, checks, and services
POS	parametric operational situation
PVT	production verification test
R&M	reliability and maintainability
RAM	reliability, availability, and maintainability
RLR	relative loss ratio

SASO	stability and support operations
SBCT	Stryker Brigade Combat Team
SEP	system evaluation plan
SME	subject-matter experts
SOSE	security operations in a stable environment
SSC	small-scale contingencies
T&E	test and evaluation
TEMP	test and evaluation master plan
TER	test evaluation report
TDP	test design plan
TOP	test operation procedures
TRADOC	Training and Doctrine Command, U.S. Army
UAV	unmanned aerial vehicles
UNKUNK	unknown unknowns

Phase I Report

Operational Test Design and Evaluation of the Interim Armored Vehicle

Executive Summary

This report provides an assessment of the U.S. Army's planned initial operational test (IOT) of the Stryker family of vehicles. Stryker is intended to provide mobility and "situation awareness" for the Interim Brigade Combat Team (IBCT). For this reason, the Army Test and Evaluation Command (ATEC) has been asked to take on the unusual responsibility of testing both the vehicle and the IBCT concept.

Building on the recommendations of an earlier National Research Council study and report (National Research Council, 1998a), the Panel on Operational Test Design and Evaluation of the Interim Armored Vehicle considers the Stryker IOT an excellent opportunity to examine how the defense community might effectively use test resources and analyze test data. The panel's judgments are based on information gathered during a series of open forums and meetings involving ATEC personnel and experts in the test and evaluation of systems. Perhaps equally important, in our view the assessment process itself has had a salutary influence on the IOT design for the IBCT/Stryker system.

We focus in this report on two aspects of the operational test design and evaluation of the Stryker: (1) the measures of performance and effectiveness used to compare the IBCT equipped with the Stryker against the baseline force, the Light Infantry Brigade (LIB), and (2) whether the current operational test design is consistent with state-of-the-art methods. Our next report will discuss combining information obtained from the

IOT with other tests, engineering judgment, experience, and the like. The panel's final report will encompass both earlier reports and any additional developments.

OVERALL TEST PLANNING

Two specific purposes of the IOT are to determine whether the IBCT/Stryker performs more effectively than the baseline force, and whether the Stryker family of vehicles meets its capability and performance requirements. Our primary recommendation is to supplement these purposes: when evaluating a large, complex, and critical weapon system such as the Stryker, operational tests should be designed, carried out, and evaluated with a view toward *improving* the capabilities and performance of the system.

MEASURES OF EFFECTIVENESS

We begin by considering the definition and analysis of measures of effectiveness (MOEs). In particular, we address problems associated with rolling up disparate MOEs into a single overall number, the use of untested or ad hoc force ratio measures, and the requirements for calibration and scaling of subjective evaluations made by subject-matter experts (SMEs). We also identify a need to develop scenario-specific MOEs for noncombat missions, and we suggest some possible candidates for these. Studying the question of whether a single measure for the "value" of situation awareness can be devised, we reached the tentative conclusion that there is no *single* appropriate MOE for this multidimensional capability. Modeling and simulation tools can be used to this end by augmenting test data during the evaluation. These tools should be also used, however, to develop a better understanding of the capabilities and limitations of the system in general and the value of situation awareness in particular.

With respect to determining critical measures of reliability and maintainability (RAM), we observe that the IOT will provide a relatively small amount of vehicle operating data (compared with that obtained in training exercises and developmental testing) and thus may not be sufficient to address all of the reliability and maintainability concerns of ATEC. This lack of useful RAM information will be exacerbated by the fact that the IOT is to be performed without using add-on armor.

For this reason, we stress that RAM data collection should be an ongoing enterprise, with failure times, failure modes, and maintenance information tracked for the entire life of the vehicle (and its parts), including data from developmental testing and training, and recorded in appropriate databases. Failure modes should be considered separately, rather than assigning a single failure rate for a vehicle using simple exponential models.

EXPERIMENTAL DESIGN

With respect to the experimental design itself, we are very concerned that observed differences will be confounded by important sources of uncontrolled variation. In particular, as pointed out in the panel's letter report (Appendix A), the current test design calls for the IBCT/Stryker trials to be run at a different time from the baseline trials. This design may confound time of year with the primary measure of interest: the difference in effectiveness between the baseline force and the IBCT/Stryker force. We therefore recommend that these events be scheduled as closely together in time as possible, and interspersed if feasible. Also, additional potential sources of confounding, including player learning and nighttime versus daytime operations, should be addressed with alternative designs. One alternative to address confounding due to player learning is to use four separate groups of players, one for each of the two opposing forces (OPFORs), one for the IBCT/Stryker, and one for the baseline system. Intergroup variability appears likely to be a lesser problem than player learning. Also, alternating teams from test replication to test replication between the two systems under test would be a reasonable way to address differences in learning, training, fatigue, and competence.

We also point out the difficulty in identifying a test design that is simultaneously "optimized" with respect to determining how various factors affect system performance for dozens of measures, and also confirming performance either against a baseline system or against a set of requirements. For example, the current test design, constructed to compare IBCT/Stryker with the baseline, is balanced for a limited number of factors. However, it does not provide as much information about the system's advantages as other approaches could. In particular, the current design allocates test samples to missions and environments in approximatley the same proportion as would be expected in field use. This precludes focusing test samples on environments in which Stryker is designed to have advantages over the baseline system, and it allocates numerous test samples to environments for which Stryker is anticipated to provide no benefits over the baseline system. This reduces the opportunity to learn the size of the benefit that Stryker provides in various environments, as well as the reasons underlying its advantages. In support of such an approach, we present a number of specific technical suggestions for test design, including making use of test design in learning and confirming stages as well as small-scale pilot tests. Staged testing, presented as an alternative to the current design, would be particularly useful in coming to grips with the difficult problem of understanding the contribution of situation awareness to system performance. For example, it would be informative to run pilot tests with the Stryker situation awareness capabilities intentionally degraded or turned off, to determine the value they provide in particular missions or scenarios.

We make technical suggestions in several areas, including statistical power calculations, identifying the appropriate test unit of analysis, combining SME ratings, aggregation, and graphical methods.

SYSTEM EVALUATION AND IMPROVEMENT

More generally, we examined the implications of this particular IOT for future tests of similar systems, particularly those that operationally interact so strongly with a novel force concept. Since the size of the operational test (i.e., number of test replications) for this complex system (or systems of systems) will be inadequate to support hypothesis tests leading to a decision on whether Stryker should be passed to full-rate production, ATEC should augment this decision with other techniques. At the very least, estimates and associated measures of precision (e.g., confidence intervals) should be reported for various MOEs. In addition, the reporting and use of numerical and graphical assessments, based on data from other tests and trials, should be explored. In general, complex systems should not be forwarded to operational testing, absent strategic considerations, until the system design is relatively mature. Forwarding an immature system to operational test is an expensive way to discover errors that could have been detected in developmental testing, and it reduces the ability of an operational test to carry out its proper function.

As pointed out in the panel's letter report (Appendix A), it is extremely important, when testing complex systems, to prepare a straw man test evaluation report (TER), as if the IOT had been completed. It should include examples of how the representative data will be analyzed, specific presentation formats (including graphs) with expected results, insights to develop from the data, draft recommendations, and so on. The content of this straw man report should be based on the experience and intuition of the analysts and what they think the results of the IOT might look like. To do this and to ensure the validity and persuasiveness of evaluations drawn from the testing, ATEC needs a cadre of statistically trained personnel with "ownership" of the design and the subsequent test and evaluation. Thus, the Department of Defense in general and ATEC in particular should give a high priority to developing a contractual relationship with leading practitioners in the fields of reliability estimation, experimental design, and data analysis to help them with future IOTs.

In summary, the panel has a substantial concern about confounding in the current test design for the IBCT/Stryker IOT that needs to be addressed. If the confounding issues were reduced or eliminated, the remainder of the test design, aside from the power calculations, has been competently developed from a statistical point of view. Furthermore, this report provides a number of evaluations and resulting conclusions and recommendations for improvement of the design, the selection and validation of MOEs, the evaluation process, and the conduct of future tests of highly complex systems. We attach greater priority to several of these recommendations and therefore highlight them here, organized by chapters to assist those interested in locating the supporting arguments.

RECOMMENDATIONS

Chapter 3

- Different MOEs should not be rolled up into a single overall number that tries to capture effectiveness or suitability.
- To help in the calibration of SMEs, each should be asked to review his or her own assessment of the Stryker IOT missions, for each scenario, immediately before he or she assesses the baseline missions (or vice versa).
- ATEC should review the opportunities and possibilities for SMEs to contribute to the collection of objective data, such as times to complete certain subtasks, distances at critical times, etc.
- ATEC should consider using two separate SME rating scales: one for "failures" and another for "successes."
- FER (and the LER when appropriate), but not the RLR, should be used as the primary mission-level MOE for analyses of engagement results.
- ATEC should use fratricide frequency and civilian casualty frequency to measure the amount of fratricide and collateral damage in a mission.

- Scenario-specific MOPs should be added for SOSE missions.
- Situation awareness should be introduced as an explicit test condition.
- RAM data collection should be an ongoing enterprise. Failure and maintenance information should be tracked on a vehicle or part/system basis for the entire life of the vehicle or part/system. Appropriate databases should be set up. This was probably not done with those Stryker vehicles already in existence, but it could be implemented for future maintenance actions on all Stryker vehicles.
- With respect to the difficulty of reaching a decision regarding reliability, given limited miles and absence of add-on-armor, weight packs should be used to provide information about the impact of additional weight on reliability.
- Failure modes should be considered separately rather than trying to develop failure rates for the entire vehicle using simple exponential models. The data reporting requirements vary depending on the failure rate function.

Chapter 4

- Given either a learning or a confirmatory objective, *ignoring various tactical considerations*, a requisite for operational testing is that it should not commence until the system design is mature.
- ATEC should consider, for future test designs, relaxing various rules of test design that it adheres to, by (a) not allocating sample size to scenarios according to the OMS/MP, but instead using principles from optimal experimental design theory to allocate sample size to scenarios, (b) testing under somewhat more extreme conditions than typically will be faced in the field, (c) using information from developmental testing to improve test design, and (d) separating the operational test into at least two stages, learning and confirmatory.
- ATEC should consider applying to future operational testing in general a two-phase test design that involves, first, learning phase studies that examine the test object under different conditions, thereby helping testers design further tests to elucidate areas of greatest uncertainty and importance, and, second, a phase involving confirmatory tests to address hypotheses concerning performance vis-à-vis a baseline system or in comparison with requirements. ATEC should consider taking advantage of this approach for the IBCT/Stryker IOT. That is, examining in the first phase IBCT/Stryker under different conditions, to assess when this system

works best, and why, and conducting a second phase to compare IBCT/Stryker to a baseline, using this confirmation experiment to support the decision to proceed to full-rate production. An important feature of the learning phase is to test with factors at high stress levels in order to develop a complete understanding of the system's capabilities and limitations.

- When specific performance or capability problems come up in the early part of operational testing, small-scale pilot tests, focused on the analysis of these problems, should be seriously considered. For example, ATEC should consider test conditions that involve using Stryker with situation awareness degraded or turned off to determine the value that it provides in particular missions.
- ATEC should eliminate from the IBCT/Stryker IOT one significant potential source of confounding, seasonal variation, in accordance with the recommendation provided earlier in the October 2002 letter report from the panel to ATEC (see Appendix A). In addition, ATEC should also seriously consider ways to reduce or eliminate possible confounding from player learning, and day/night imbalance.

Chapter 5

- The IOT provides little vehicle operating data and thus may not be sufficient to address all of the reliability and maintainability concerns of ATEC. This highlights the need for improved data collection regarding vehicle usage. In particular, data should be maintained for each vehicle over that vehicle's entire life, including training, testing, and ultimately field use; data should also be gathered separately for different failure modes.
- The panel reaffirms the recommendation of the 1998 NRC panel that more use should be made of estimates and associated measures of precision (or confidence intervals) in addition to significance tests, because the former enable the judging of the practical significance of observed effects.

Chapter 6

- Operational tests should not be strongly geared toward estimation of system suitability, since they cannot be expected to run long enough to estimate fatigue life, estimate repair and replacement times, identify failure modes, etc. Therefore, developmental testing should give greater priority to measurement of system (operational) suitability and should be structured to provide its test events with greater operational realism.
- In general, complex systems should not be forwarded to operational

testing, in the absence of strategic considerations, until the system design is relatively mature. Forwarding an immature system to operational test is an expensive way to discover errors that could have been detected in developmental testing, and it reduces the ability of an operational test to carry out its proper function. System maturation should be expedited through previous testing that incorporates various aspects of operational realism in addition to the usual developmental testing.

- Because it is not yet clear that the test design and the subsequent test analysis have been linked, ATEC should prepare a straw man test evaluation report in advance of test design, as recommended in the panel's October 2002 letter to ATEC (see Appendix A).
- The goals of the initial operational test need to be more clearly specified. Two important types of goals for operational test are learning about system performance and confirming system performance in comparison to requirements and in comparison to the performance of baseline systems. These two different types of goals argue for different stages of operational test. Furthermore, to improve test designs that address these different types of goals, information from previous stages of system development need to be utilized.

Finally, we wish to make clear that the panel was constituted to address the statistical questions raised by the selection of measures of performance and measures of effectiveness, and the selection of an experimental design, given the need to evaluate Stryker and the IBCT in scenarios identified in the OMS/MP. A number of other important issues (about which the panel provides some commentary) lie outside the panel's charge and expertise. These include an assessment of (a) the selection of the baseline system to compare with Stryker, (b) the problems raised by the simultaneous evaluation of the Stryker vehicle and the IBCT system that incorporates it, (c) whether the operational test can definitively answer specific tactical questions, such as the degree to which the increased vulnerability of Stryker is offset by the availability of greater situational awareness, and (d) whether or not scenarios to be acted out by OPFOR represent a legitimate test suite. Let us elaborate each of these ancillary but important issues.

The first is whether the current choice of a baseline system (or multiple baselines) is best from a military point of view, including whether a baseline system could have been tested taking advantage of the IBCT infrastructure, to help understand the value of Stryker without the IBCT system. It does not seem to be necessary to require that only a system that could be transported as quickly as Stryker could serve as a baseline for comparison.

The second issue (related to the first) is the extent to which the current test provides information not only about comparison of the IBCT/Stryker system with a baseline system, but also about comparison of the Stryker suite of vehicles with those used in the baseline. For example, how much more or less maneuverable is Stryker in rural versus urban terrain and what impact does that have on its utility in those environments? These questions require considerable military expertise to address.

The third issue is whether the current operational test design can provide adequate information on how to tactically employ the IBCT/Stryker system. For example, how should the greater situational awareness be taken advantage of, and how should the greater situational awareness be balanced against greater vulnerability for various types of environments and against various threats? Clearly, this issue is not fundamentally a technical statistical one, but is rather an essential feature of scenario design that the panel was not constituted to evaluate.

The final issue (related to the third) is whether the various missions, types of terrain, and intensity of conflict are the correct choices for operational testing to support the decision on whether to pass Stryker to full-rate production. One can imagine other missions, types of terrain, intensities, and other factors that are not varied in the current test design that might have an impact on the performance of Stryker, the baseline system, or both. These factors include temperature, precipitation, the density of buildings, the height of buildings, types of roads, etc. Moreover, there are the serious problems raised by the unavailability of add-on armor for the early stages of the operational test. The panel has been obligated to take the OMS/MP as given, but it is not clear whether additional factors that might have an important impact on performance should have been included as test factors. All of these issues are raised here in order to emphasize their importance and worthiness for consideration by other groups better constituted to address them.

Thus, the panel wishes to make very clear that this assessment of the operational test as currently designed reflects only its statistical merits. It is certainly possible that the IBCT/Stryker operational test may be deficient in other respects, some of them listed above, that may subordinate the statistical aspects of the test. Even if the statistical issues addressed in this report were to be mitigated, we cannot determine whether the resulting operational test design would be fully informative as to whether Stryker should be promoted to full-rate production.

1

Introduction

This report provides an assessment of the U.S. Army's planned initial operational test of the Stryker family of vehicles. It focuses on two aspects of the test design and evaluation: (1) the measures of performance and effectiveness used to compare the force equipped with the Stryker with a baseline force and (2) whether the current operational test design is consistent with state-of-the-art methods.

ARMY'S NEED FOR AN INTERIM ARMORED VEHICLE (STRYKER)

The United States Army anticipates increases in the number and types of asymmetric threats and will be required to support an expanding variety of missions (including military operations in urban terrain and operations other than war) that demand an effective combination of rapid deployability, information superiority, and coordination of awareness and action. In order to respond to these threats and mission requirements, the Army has identified the need for a future combat system that leverages the capabilities of advancing technologies in such areas as vehicle power, sensors, weaponry, and information gathering and sharing. It will take years to develop and integrate these technologies into weapon systems that meet the needs of the Army of the future. The Army has therefore established a three-pronged plan to guide the transition of its weapons and forces, as illustrated in Figure 1-1.

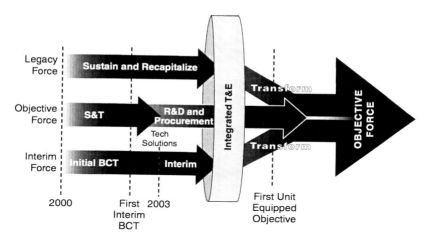

FIGURE 1-1 Army plans for transition to the Objective Force equipped with the future combat system.
Acronyms: BCT, brigade combat team; R&D, research and development; S&T, science and technology; T&E, test and evaluation.
SOURCE: ATEC briefing to the panel.

The Army intends to use this transition plan to sustain and upgrade (but not expand) its existing weapon systems, which are characterized as its Legacy Force. Heavy armored units, a key element of the Legacy Force, are not, however, adequate to address the challenge of rapid deployability around the globe.

An immediate and urgent need exists for an air transportable Interim Force capable of deployment to anywhere on the globe in a combat-ready configuration. Until the future combat system is developed and fully fielded to support the Objective Force (the Army of the future), the Army intends to rely on an Interim Force, the critical warfighting component of which is the Interim Brigade Combat Team (IBCT). The mobility of the IBCT is to be provided by the Stryker vehicle (until recently referred to as the interim armored vehicle).

Stryker Configurations

The range of tasks to be accomplished by the IBCT calls for the Stryker to be not just a single vehicle, but a family of vehicles that are air transportable, are capable of immediate employment upon arrival in the area of operations, and have a great degree of commonality in order to decrease its logistical "footprint." Table 1-1 identifies the various Stryker vehicle configurations (as identified in the Stryker system evaluation plan), the key government-furnished equipment (GFE) items that will be integrated into the configuration, and the role of each configuration.

Stryker Capabilities

The Army has identified two different but clearly dependent capability requirements for the Stryker-supported IBCT: operational capabilities for the IBCT force that will rely on the Stryker, and system capabilities for the Stryker family of vehicles.

IBCT Operational Capabilities

The Army's Operational Requirements Document for the Stryker (ORD, 2000) defines the following top-level requirement for the IBCT:

> The IBCT is a full spectrum, combat force. It has utility, confirmed through extensive analysis, in all operational environments against all projected future threats, but it is designed and optimized primarily for employment in small scale contingency (SSC) operations in complex and urban terrain, confronting low-end and mid-range threats that may employ both conventional and asymmetric capabilities. The IBCT deploys very rapidly, executes early entry, and conducts effective combat operations immediately on arrival to prevent, contain, stabilize, or resolve a conflict through shaping and decisive operations (section 1.a.(3)) . . . As a full spectrum combat force, the IBCT is capable of conducting all major doctrinal operations including offensive, defensive, stability, and support actions. . . . Properly integrated through a mobile robust C4ISR network, these core capabilities compensate for platform limitations that may exist in the close fight, leading to enhanced force effectiveness (section 1.a.(4)).

System Capabilities

Each configuration of the Stryker vehicle must be properly integrated with sensing, information processing, communications, weapons, and other essential GFE that has been developed independently of the Stryker. The Army notes that the Stryker-GFE system must provide a particular capability, termed *situation awareness,* to offset the above-mentioned platform limitations:

TABLE 1-1 Stryker Configurations

Configuration Name	Government-Furnished Equipment	Role
Infantry carrier vehicle	Force XXI Battle Command Brigade and Below (FBCB2), Enhanced Position Location and Reporting System (EPLRS), Global Positioning System (GPS), Thermal Weapon Sight (TWS-H)	Carry and protect a nine-man infantry squad and a crew of two personnel.
Mortar carrier	M121 120-mm mortar, XM95 Mortar Fire Control System	Provide fire support to maneuver forces.
Antitank guided missile vehicle	TOW II missile system, TOW acquisition system	Defeat armored threats.
Reconnaissance vehicle	Long Range Advanced Scout Surveillance System (LRAS3)	Enable scouts to perform reconnaissance and surveillance.
Fire support vehicle	Mission Equipment Package (MEP), Handheld Terminal Unit (HTU), Common Hardware-Software – Lightweight Computer Unit (CHS-LCU)	Provide automation-enhanced target acquisition, identification, and detection; communicate fire support information.

The IBCT will offset the lethality and survivability limitations of its platforms through the holistic integration of all other force capabilities, particularly the internetted actions of the combined arms company teams. The mounted systems equipped with Force XXI Battle Command, Brigade and Below (FBCB2) and other enhancements provide the IBCT a larger internetted web of situational awareness extending throughout the IBCT area of operations. The synergistic effects achieved by internetting highly trained soldiers and leaders with platforms and organizational design enable the force to avoid surprise, develop rapid decisions, control the time and place to engage in combat, conduct precision maneuver, shape the battlespace with precision fires and effects, and achieve decisive outcomes. (ORD section 1.a.(7)).

The Stryker ORD specifies several performance requirements that ap-

TABLE 1-1 Continued

Configuration Name	Government-Furnished Equipment	Role
Engineer squad vehicle	FBCB2, EPLRS, GPS, TWS-H	Neutralize and mark obstacles, detect mines, transport engineering squad.
Commander's vehicle	All Source Analysis System (ASAS), Advanced Field Artillery Tactical Data System Command, Control, Communications and Computer (AFATDS C4).	Provide command, control, communications, and computer attributes to enable commanders to direct the battle.
Medical evacuation vehicle	MC-4 Medic's Aide	Enable recovery and evacuation of casualties.
Nuclear, biological, chemical (NBC) reconnaissance vehicle	NBC Sensor Suite	Perform reconnaissance missions in NBC environment; detect NBC conditions.
Mobile gun system	Eyesafe Laser Range Finder (ELRF)	Provide weapons fire to support assaulting infantry.

ply across configurations of the Stryker system. Key performance requirements, defined in more detail in the ORD, are that the Stryker vehicles must:

- maximize commonality of components and subcomponents across configurations;
- possess an "internetted interoperable capability" that enables it to host and integrate existing and planned Army command, control, communications, computers, intelligence, surveillance, and reconnaissance (C4ISR) systems;
- be transportable in a C-130 aircraft;

- have the ability to operate effectively 24 hours per day, including at night, in inclement weather, and during other periods of limited visibility in hot, temperate, and cold climates;
- be mobile, demonstrable in terms of specified hard surface speeds, cross-country mobility, cruising range, ability to transverse obstacles and gaps, and maneuverability;
- possess the capability of sustainability, indicated by specified abilities to tow and be towed, to be started with assistance, to be refueled rapidly, and to provide auxiliary power in the event of loss of primary power;
- be survivable, as evidenced by the ability to achieve specified acceleration, to provide protection to its crew and equipment, to accept add-on armor, to mount weapons, and to rapidly self-obscure;
- permit the survivability of its crew in external environments with nuclear, biological, and chemical contamination, by providing warning or protection;
- possess the capability of lethality, demonstrated by its ability to inflict lethal damage on opposing forces and weapon systems;
- satisfy specified requirements for logistics and readiness, which contribute to its fundamental requirement to maintain the force in the field;
- be transportable by air and by rail;
- operate reliably (i.e., without critical failures) for a specified period of time and be maintainable within a specified period of time when failures do occur.

The Army Test and Evaluation Command (ATEC) has been assigned the mission of testing, under operationally realistic conditions, and evaluating the extent to which the IBCT force equipped with the Stryker system (IBCT/Stryker) meets its requirements, compared with a baseline force, the current Light Infantry Brigade (LIB), which will be augmented with transportation assets appropriate to assigned missions.

Although we have major concerns about the appropriateness of using the LIB as an alternative comparison force, because our primary responsibility is to address broader statistical and test design issues, we have taken this choice of baseline to be a firm constraint.

PANEL CHARGE

The Stryker will soon be entering an 18-month period of operational test and evaluation to determine whether it is effective and suitable to enter into full-rate production, and how the vehicles already purchased can best be used to address Army needs. Typical of the operational test and evaluation of a major defense acquisition program, the test and evaluation of the Stryker is an extremely complicated undertaking involving several separate test events, the use of modeling and simulation, and a wide variety of requirements that need to be satisfied.

Reacting to a very high level of congressional interest in the Stryker program, ATEC must develop and use an evaluation approach that applies statistical rigor to determining the contribution of the Stryker to the IBCT mission, as well as the effectiveness of the IBCT itself. Affirming the value of obtaining an independent assessment of its approach, and desiring assistance in developing innovative measures of effectiveness, ATEC requested that the National Research Council's (NRC) Committee on National Statistics convene a panel of experts to examine its plans for the operational test design and subsequent test and evaluation for the IBCT/Stryker. This resulted in the formation of the Panel on Operational Test Design and Evaluation of the Interim Armored Vehicle. The panel was specifically charged to examine three aspects of the operational test and evaluation of the IBCT/Stryker:

1. the measures of performance and effectiveness used to compare the IBCT force equipped with the Stryker system to the baseline, the LIB, and the measures used to assess the extent to which the Stryker system meets its requirements;

2. the design of the operational test, to determine the extent to which the design is consistent with state-of-the art methods in statistical experimental design; and

3. the applicability of combining information models, as well as of combining information from testing and field use of related systems and from developmental test results for the Stryker, with operational test results for the Stryker.

The panel was also asked to identify alternative measures (e.g., of situation awareness) and experimental designs that could better reflect the advantages and disadvantages of the IBCT force equipped with the Stryker system relative to the LIB force. In addition, the panel was asked to address the use of modeling and simulation as part of the program evaluation and the analysis strategy proposed for the evaluation of the IBCT/Stryker.

PANEL APPROACH

In its 1998 report *Statistics, Testing, and Defense Acquisition: New Approaches and Methodological Improvements*, the NRC Committee on National Statistics' Panel on Testing and Evaluating Defense Systems established broad perspectives and fundamental principles applicable to the examination of statistical aspects of operational testing (National Research Council, 1998a). Our panel has adopted the findings, conclusions, and recommendations of that report as a key starting point for our deliberations.

We also reviewed in detail all key government documents pertaining to the operational testing of the IBCT/Stryker, including:

- the Operational Requirements Document,
- the System Evaluation Plan,
- the Test and Evaluation Master Plan,
- the Organizational and Operational Description,
- the Failure Definition and Scoring Document,
- the Mission Needs Statement,
- the Operational Mode Summary and Mission Profile, and
- sample Operational Orders applicable to operational tests.

With the cooperation of the management and staff of ATEC, the panel conducted two forums and two subgroup meetings at which ATEC staff presented, in response to panel queries: descriptions of measures of effectiveness, suitability, and survivability under consideration for the initial operational test; details of the proposed experimental design; planned use of modeling and simulation; and planned methods of analysis of data that will result from the testing. At these forums, panel members and ATEC staff engaged in interactive discussion of proposed and alternative measures, test designs, and analytical methods. At the invitation of the panel, two forums—one on measures and one on test designs—were attended by representatives from the Office of the Director, Operational Test and Evaluation; the Institute for Defense Analysis; the U.S. Government Accounting Office; and the U.S. Military Academy at West Point. Beyond the specific recommendations and conclusions presented here, it is our view that the open and pointed discussions have created a process that in itself has had a salutary influence on the decision making and design for the testing of the Stryker system of vehicles and the IBCT.

This report summarizes the panel's assessment regarding (1) the measures of performance and effectiveness used to compare the IBCT force equipped with the Stryker system with the baseline force and the extent to which the Stryker system meets its requirements and (2) the experimental design of the operational test. This report also addresses measures for situation awareness, alternative measures for force effectiveness, analysis strategies, and some issues pertaining to modeling and simulation. Box 1-1 presents a number of terms used in operational testing.

After additional forums and deliberations, the panel intends to prepare a second report that addresses the applicability of combining information from other sources with that from the IBCT/Stryker initial operational test. Those sources include developmental tests for the Stryker, testing and field experience with related systems, and modeling and simulation. Our

---

BOX 1-1 Acronyms Used in Operational Testing

ABCS	Army Battle Command Systems
AoA	Add-on armor
ATEC	Army Test and Evaluation Command
ATIRS	Army test incident reporting system
BLUFOR	Blue force
BRL/AMSAA	Ballistics Research Laboratories/U.S. Army Materiel Systems Analysis Activity
C4ISR	Command, control, communications, computers, intelligence, surveillance, and reconnaissance
CAA	Center for Army Analysis
CCTT	Close combat tactical training
CF	Casualty frequency
DoD	Department of Defense
DT	Developmental test
DTC	Developmental Test and Command
DTP	Detailed test plan
DOT&E	Director, Operational Test and Evaluation
FBCB2	Force XXI Battle Command Brigade and Below
FCS	Future combat system
FDSC	Failure definition and scoring criteria
FER	Force exchange ratio

continued on next page

BOX 1-1 Continued

FF	Fratricide frequency
FLIR	Forward-looking infrared systems
GFE	Government-furnished equipment
IAV	Interim armored vehicle
IBCT	Interim Brigade Combat Team
ICV	Infantry carrier vehicle
IOT	Initial operational test
LER	Loss exchange ratio
LIB	Light Infantry Brigade
MEP	Mission Equipment Package
MIB	Mechanized Infantry Brigade
MIL-STD	Military standard
MOE	Measure of effectiveness
MOP	Measure of performance
MR	Maintenance ratio
MTTR	Mean time to repair
NBC	Nuclear/biological/chemical
OMS/MP	Operational mission summary/mission profile
OPFOR	Opposition force
ORD	Operational requirements document
OT	Operational tests
PMCS	Preventive maintenance, checks, and services
PVT	Product verification test
R&M	Reliability and maintainability
RAM	Reliability, availability, and maintainability
RLR	Relative loss ratio
SASO	Stability and support operations
SEP	System evaluation plan
SME	Subject-matter expert
SOSE	Security operations in a stability environment
SSC	Small-scale contingency
T&E	Test and evaluation
TEMP	Test and evaluation master plan
TER	Test evaluation report
TDP	Test design plan
TOP	Test operation procedures
TRAC-FLVN	U.S. Army Training and Doctrine Command Analysis Center, Fort Leavenworth, Kansas
TRADOC	U.S. Army Training and Doctrine Command
UAV	Unmanned aerial vehicles

final report will incorporate both Phase I and II reports and any additional developments.

Finally, we wish to make clear that the panel was constituted to address the statistical questions raised by the selection of measures of performance and measures of effectiveness, and the selection of an experimental design, given the need to evaluate Stryker and the IBCT in scenarios identified in the OMS/MP. A number of other important issues (about which the panel provides some commentary) lie outside the panel's charge and expertise. These include an assessment of (a) the selection of the baseline system to compare with Stryker, (b) the problems raised by the simultaneous evaluation of the Stryker vehicle and the IBCT system that incorporates it, (c) whether the operational test can definitively answer specific tactical questions, such as the degree to which the increased vulnerability of Stryker is offset by the availability of greater situational awareness, and (d) whether or not scenarios to be acted out by OPFOR represent a legitimate test suite. Let us elaborate each of these ancillary but important issues.

The first is whether the current choice of a baseline system (or multiple baselines) is best from a military point of view, including whether a baseline system could have been tested taking advantage of the IBCT infrastructure, to help understand the value of Stryker without the IBCT system). It does not seem to be necessary to require that only a system that could be transported as quickly as Stryker could serve as a baseline for comparison.

The second issue (related to the first) is the extent to which the current test provides information not only about comparison of the IBCT/Stryker system with a baseline system, but also about comparison of the Stryker suite of vehicles with those used in the baseline. For example, how much more or less maneuverable is Stryker in rural versus urban terrain and what impact does that have on its utility in those environments? These questions require considerable military expertise to address.

The third issue is whether the current operational test design can provide adequate information on how to tactically employ the IBCT/Stryker system. For example, how should the greater situational awareness be taken advantage of, and how should the greater situational awareness be balanced against greater vulnerability for various types of environments and against various threats? Clearly, this issue is not fundamentally a technical statistical one, but is rather an essential feature of scenario design that the panel was not constituted to evaluate.

The final issue (related to the third) is whether the various missions, types of terrain, and intensity of conflict are the correct choices for operational testing to support the decision on whether to pass Stryker to full-rate production. One can imagine other missions, types of terrain, intensities, and other factors that are not varied in the current test design that might have an impact on the performance of Stryker, the baseline system, or both. These factors include temperature, precipitation, the density of buildings, the height of buildings, types of roads, etc. Moreover, there are the serious problems raised by the unavailability of add-on armor for the early stages of the operational test. The panel has been obligated to take the OMS/MP as given, but it is not clear whether additional factors that might have an important impact on performance should have been included as test factors. All of these issues are raised here in order to emphasize their importance and worthiness for consideration by other groups better constituted to address them.

Thus, the panel wishes to make very clear that this assessment of the operational test as currently designed reflects only its statistical merits. It is certainly possible that the IBCT/Stryker operational test may be deficient in other respects, some of them listed above, that may subordinate the statistical aspects of the test. Even if the statistical issues addressed in this report were to be mitigated, we cannot determine whether the resulting operational test design would be fully informative as to whether Stryker should be promoted to full-rate production.

2

Test Process

The Army's interim armored combat vehicle, now called the Stryker, is in the latter stages of development. As is the case in all major acquisitions, it is necessary for the Army to subject the vehicle to a set of tests and evaluations to be sure it understands what it is buying and how well it works in the hands of the users. The Army Test and Evaluation Command (ATEC) has been charged to do these tests and evaluations. A key element in this series of tests is the operational testing, commencing with the initial operational test (IOT).

The basic thought in the development of any operational test plan is to test the equipment in an environment as similar as possible to the environment in which the equipment will actually operate. For combat vehicles such as the Stryker, the standard practice is to create combat situations similar to those in which the test vehicle would be expected to perform. The system is then inserted into the combat situations with trained operators and with an opposition force (OPFOR) of the type expected. Preplanned scenarios and training schedules for the players in the test are developed, the nature of the force in which the test vehicle will be embedded is identified, and the test plans are developed by ATEC.

Testing and evaluation of the Stryker is especially challenging, because several issues must be addressed together:

- To what extent does the Stryker—in various configurations, equipped with integrated government-furnished equipment (GFE)—meet (or fail to meet) its requirements (e.g., for suitability and survivability)?

• How effective is the Interim Brigade Combat Team (IBCT), equipped with the Stryker system, and how does its effectiveness compare with that of a baseline force?
• What factors (in the forces and in the systems) account for successes and failures in performance and for any performance differences between the forces and the systems?

Thus, a primary objective of the IOT is to compare an organization that includes the Stryker with a baseline organization that does not include the Stryker. This makes the evaluation of the data particularly important and challenging, because the effects of the differences in organizations will be confounded with the differences in their supporting equipment systems.

The planning for the test of the IBCT/Stryker is particularly difficult because of the complex interactions among these issues, the varying missions in which the IBCT/Stryker will be tested, the number of variants of the vehicle itself, time and budget constraints (which affect the feasible size and length of tests), uncertainty about the characteristics of the planned add-on armor, and the times of year at which the test will be run.

Factors that must be considered in the evaluation of the test data are:

1. Modeling and simulation using the results of the live test, accounting for the uncertainty due to small sample size in the live tests.
2. The incorporation of developmental test data, manufacturer test data, and historical data in the evaluation.
3. Extrapolation of IOT field data to higher echelons that, due to resource constraints, will not be tested by a live, representative force in the Stryker IOT. In particular, one of the three companies in the battalion that will be played in the Stryker IOT is notional (i.e., its communications, disposition, and effects will be simulated); the battalion headquarters has no other companies to worry about; and the brigade headquarters is played as a "white force" (neutral entity that directs and monitors operations).
4. The relative weight to give to "hard" instrument-gathered data vis-à-vis the observations and judgments of subject-matter experts (SMEs).

OVERALL TESTING AND EVALUATION PLAN

Two organizations, the government and the contractors that build the systems, are involved in the test and evaluation of Army systems. The

Army's Developmental Test Command (within the ATEC organization) conducts, with contractor support, the production verification test. The purpose of this test is to ensure that the system, as manufactured, meets all of the specifications given in the contract. This information can be valuable in the design and evaluation of results of subsequent developmental tests, particularly the testing of reliability, availability, and maintainability (RAM).

Within the Army, responsibility for test and evaluation is given to ATEC. When ATEC is assigned the responsibility for performing test and evaluation for a given system, several documents are developed:

- the test and evaluation master plan,
- the test design plan,
- the detailed test plan,
- the system evaluation plan, and
- the failure definition and scoring criteria.

Testers perform a developmental test on the early production items in order to verify that the specifications have been met or exceeded (e.g., a confirmation, by noncontractor personnel, of the product verification test results on delivered systems). Following the developmental test, ATEC designs and executes one or more operational tests, commencing with the IOT. Modeling and simulation are often used to assist in test design, to verify test results, and to add information that cannot be obtained from the IOT.

EVALUATION OF THE DATA

When the results of the product verification test, the developmental test, the initial operational test, the modeling and simulation, and the history of use have been gathered, ATEC is responsible for compiling all data relevant to the system into a final evaluation report. The Director, Operational Test and Evaluation, approves the ATEC IOT event design plan and conducts an independent evaluation. As noted above, IBCT/Stryker testing will address two fundamental questions: (1) To what extent does the Stryker system (i.e., integration of the Stryker vehicle and its GFE in various configurations) meet its requirements? (2) How well does the IBCT force, equipped with the Stryker system, perform and meet its requirements, compared with the baseline Light Infantry Brigade (LIB) force? Evaluators will also assess the ways in which the IBCT force employs the Stryker and the extent to which the Stryker GFE provides situation awareness to the IBCT. They will also use the test data to help develop an understanding of why the IBCT and Stryker perform as well (or poorly) as they do.

The ATEC evaluator is often asked about the most effective way to employ the system. If the test has been designed properly, extrapolation from the test data can often shed light on this question. In tests like those planned for the IBCT/Stryker, the judgment of the SMEs is highly valuable. In the design, the SMEs may be asked to recommend, after early trials of the Stryker, changes to make the force more effective. This process of testing, recommending improvements, and implementing the recommendations can be done iteratively. Clearly, although the baseline trials can provide helpful insights, they are not intended primarily to support this kind of analysis.

With the outcome of each test event recorded and with the aid of modeling, the evaluator will also extrapolate from the outcome of the IOT trials to what the outcome would have been under different circumstances. This extrapolation can involve the expected outcomes at different locations, with different force sizes, or with a full brigade being present, for example.

TEST PROCESS

Scripting

In the test design documents, each activity is scripted, or planned in advance. The IOT consists of two sets of operational trials, currently planned to be separated by approximately three weeks: one using the IBCT/Stryker, the other the baseline LIB. Each trial is scheduled to have a nine-day duration, incorporating three types of mission events (raid, perimeter defense, and security operations in a stability environment) scripted during the nine days. The scripting indicates where and when each of these events in the test period occurs. It also establishes starting and stopping criteria for each event. There will be three separate nine-day trials using the IBCT test force and three separate nine-day trials using the LIB baseline force.

Integrated Logistics Support

Logistics is always a consideration during an IOT. It will be especially important in Stryker, since the length of a trial (days) is longer than for typical weapon system IOT trials (hours). The supporting unit will be assigned in advance, and its actions will be controlled. It will be predetermined whether a unit can continue based on the logistical problems encountered. The handling of repairs and replacements will be scripted.

The role of the contractor in logistics support is always a key issue: contractors often maintain systems during introduction to a force, and both the level of training of Army maintenance personnel and the extent of contractor involvement in maintenance can affect force and system performance during operational testing. The contractor will not be present during actual combat, so it could be argued that the contractor should not be permitted in the areas reserved for the IOT. A counterargument is that the IOT can represent an opportunity for the contractor to learn where and how system failures occur in a combat environment.

Safety

A safety officer, present at all times, attempts to ensure that safety rules are followed and is allowed to stop the trial if it becomes apparent that an unsafe condition exists.

Constraints on Test and Evaluation

The test and evaluation design and execution are greatly influenced by constraints on time, money, availability of trained participants, and availability of test vehicles, as well as by demands by the contractor, the project manager, the director of operational test and evaluation, and Congress. In the IBCT/Stryker IOT, the time constraint is especially critical. The availability of test units, test assets, test players, and test sites has created constraints on test design; implications of these constraints are discussed later in this report. One key constraint is the selection of the baseline force (LIB) and its equipment. We note that there are alternative baselines (e.g., the Mechanized Infantry Brigade and variations to the baseline equipment configurations) that could have been selected for the Stryker IOT but consider it beyond the scope of the panel's charge to assess the choice of baseline.

One possibility that might be considered by ATEC would be to have the subject-matter experts also tasked to identify test results that might have been affected had the baseline force been different. Although this might involve more speculation than would be typical for SME's given their training, their responses could provide (with suitable caveats) valuable insights.

One salient example of the effects of resource constraints on the Stryker IOT is the limited number of options available to test the situation awareness features of the Stryker's C4ISR (command, control, communications, computers, intelligence, surveillance, and reconnaissance). The evaluation of the C4ISR and the ensuing situation awareness is difficult. If time would permit, it would be valuable to run one full trial with complete information and communication and a matched trial with the information and the transmission thereof degraded. It is unlikely that this will be feasible in the Stryker IOT. In fact, it will be feasible to do only a few of the possible treatment combinations needed to consider the quality of the intelligence, the countermeasures against it, the quality of transmission, and how much information should be given to whom.

CURRENT STATISTICAL DESIGN[1]

The IBCT/Stryker IOT will be conducted using two live companies operating simultaneously with a simulated company. These companies will carry out three types of missions: raid, perimeter defense, and security operations in a stable environment. The stated objective of the operational test is to compare Stryker-equipped companies with a baseline of light infantry companies for these three types of missions.

The operational test consists of three nine-day scenarios of seven missions per scenario for each of two live companies, generating a total of 42 missions, 21 for each company, carried out in 27 days. These missions are to be carried out by both the IBCT/Stryker and the baseline force/system. We have been informed that only one force (IBCT or the baseline) can carry out these missions at Fort Knox at one time, so two separate blocks of

---

[1] The following material is taken from a slide presentation to the panel: April 15, 2002 (U.S. Department of Defense, 2002a). A number of details, for example about the treatment of failed equipment and simulated casualties, are omitted in this very brief design summary.

27 days have been reserved for testing there. The baseline LIB portion of the test will be conducted first, followed by the IBCT/Stryker portion.

ATEC has identified the four design variables to be controlled during the operational test: *mission type* (raid, perimeter defense, security operations in a stable environment), *terrain* (urban, rural), *time of day* (day, night), and *opposition force intensity* (low—civilians and partisans; medium—civilians, partisans, paramilitary units; and high—civilians, partisans, paramilitary units, and conventional units). The *scenario* (a scripted description of what the OPFOR will do, the objectives and tasks for test units, etc.) for each test is also controlled for and is essentially a replication, since both the IBCT and the baseline force will execute the same scenarios. The panel has commented in our October 2002 letter report (Appendix A) that the use of the same OPFOR for both the IBCT and the baseline trials (though in different roles), which are conducted in sequence, will introduce a learning effect. We suggested in that letter that interspersing the trials for the two forces could, if feasible, reduce or eliminate that confounding effect.

ATEC has conducted previous analyses that demonstrate that a test sample size of 36 missions for the IBCT/Stryker and for the baseline would provide acceptable statistical power for overall comparisons between them and for some more focused comparisons, for example, in urban environments. We comment on these power calculations in Chapter 4.

The current design has the structure shown in Table 2-1. The variable "time of day," which refers to whether the mission is mainly carried out during daylight or nighttime, is not explicitly mentioned in the design matrix. Although we assume that efforts will be made in real time, opportunistically, to begin missions so that a roughly constant percentage of test events by mission type, terrain, and intensity, and for both the IBCT/Stryker and the baseline companies, are carried out during daylight and nighttime, time of day should be formalized as a test factor.

It is also our understanding that, except for the allocation of the six extra missions, ATEC considers it infeasible at this point to modify the general layout of the design matrix shown in the table.

TABLE 2-1 Current Design of the IBCT/Stryker Initial Operational Test

Scenario	Scenario 1						Scenario 2						Scenario 3						Total
Company-pair	B-pair[a]			C-pair[a]			B-pair[a]			C-pair[a]			B-pair[a]			C-pair[a]			
Mission[b]	S	R	P	S	R	P	S	R	P	S	R	P	S	R	P	S	R	P	
Terrain-Intensity:																			
Rural low	1											1	1					1	6
Rural med		1		1				1							1	1			6
Rural high			1		1				1		1			1			1		6
Urban low	1		1	1			1			1					1	1			6
Urban med	1				1		1					1	1					1	6
Urban high		1				1		1		1					1		1		6
Medium[c]	1			1			1			1			1			1			6
Cells per column	3	2	2	3	2	2	3	2	2	3	2	2	3	2	2	3	2	2	42

[a]B-pair and C-pair are nominal designators for the two pairs of companies in a given IBCT or baseline test.
[b]S stands for security operations in a stable environment, R stands for raid, and P stands for perimeter defense.
[c]These six extra missions were added to provide additional power for certain comparisons, but they have been suggested by ATEC for use in examining additional issues, specifically problems raised when the global positioning satellite capability of Stryker is jammed.

3

Test Measures

The Interim Brigade Combat Team (IBCT) equipped with the Stryker is intended to provide more combat capability than the current Light Infantry Brigade (LIB) and to be significantly more strategically deployable than a heavy Mechanized Infantry Brigade (MIB). It is anticipated that the IBCT will be used in at least two roles:

1. as part of an early entry combat capability against armed threats in small-scale contingencies (SSC). These IBCT engagements are likely to be against comparable forces—forces that can inflict meaningful casualties on each other.

2. in stability and support operations against significantly smaller and less capable adversaries than anticipated in SSC. The Stryker system evaluation plan (SEP) uses the term security operations in a stability environment (SOSE); that term will be used here.

The IBCT/Stryker initial operational test (IOT) will include elements of both types of IBCT missions to address many of the issues described in the Stryker SEP. This chapter provides an assessment of ATEC's plans for measures to use in analyzing results of the IOT. We begin by offering some definitions and general information about measures as background for specific comments in subsequent sections.

INTRODUCTION TO MEASURES

Using the IBCT and the IOT as context, the following definitions are used as a basis for subsequent discussions.

The *objective* of the IBCT is synonymous with the mission it is assigned to perform. For example:

- "Attack to seize and secure the opposition force's (OPFOR) defended position" (SSC mission)
- "Defend the perimeter around . . . for x hours" (SSC mission)
- "Provide area presence to . . . " (SOSE mission)

Objectives will clearly vary at different levels in the IBCT organization (brigade, battalion, company, platoon), and several objectives may exist at one level and may in fact conflict (e.g., "Attack to seize the position and minimize friendly casualties").

Effectiveness is the extent to which the objectives of the IBCT in a mission are attained. *Performance* is the extent to which the IBCT demonstrates a capability needed to fulfill its missions effectively. Thus, performance could include, for example, the Stryker vehicle's survivability, reliability, and lethality; the IBCT's C4ISR (command, control communications, computers, intelligence, surveillance, and reconnaissance); and situation awareness, among other things.

A measure of performance (MOP) is a metric that describes the amount (or level) of a performance capability that exists in the IBCT or some of its systems. A measure of effectiveness (MOE) is a quantitative index that indicates the degree to which a mission objective of the IBCT is attained. Often many MOEs are used in an analysis because the mission may have multiple objectives or, more likely, there is a single objective with more than one MOE. For example, in a perimeter defense mission, these may include the probability that no penetration occurs, the expected value of the time until a penetration occurs, and the expected value of the number of friendly casualties, all of which are of interest to the analyst. For the IBCT IOT, mission-level MOEs can provide useful information to:

1. evaluate how well a particular mission or operation was (or will be) performed. Given appropriate data collection, they provide an objective and quantitative means of indicating to appropriate decision makers the degree of mission accomplishment;

2. provide a means of quantitatively comparing alternative forces (IBCT versus LIB); and

3. provide a means of determining the contribution of various incommensurate IBCT performance capabilities (survivability, lethality, C4ISR, etc.) to mission success (if they are varied during experiments) and therefore information about the utility of changing the level of particular capabilities.

Although numerical values of mission-level MOEs provide quantitative information about the degree of mission success, the analysis of operational test results should also be a diagnostic process, involving the use of various MOEs, MOPs, and other information to determine why certain mission results occurred. Using only summary MOE values as a rationale for decision recommendations (e.g., select A over B because $MOE_A = 3.2 > MOE_B = 2.9$) can lead to a tyranny of numbers, in which precisely stated values can be used to reach inappropriate decisions. The most important role of the analyst is to develop a causal understanding of the various factors (force size, force design, tactics, specific performance capabilities, environmental conditions, etc.) that appear to drive mission results and to report on these as well as highlight potential problem areas.

Much has been written about pitfalls and caveats in developing and using MOEs in military analyses. We mention two here because of their relevance to MOEs and analysis concepts presented in the IBCT/Stryker test and evaluation master plan (TEMP) documentation.

1. As noted above, multiple MOEs may be used to describe how well a specific mission was accomplished. Some analysts often combine these into a single overall number for presentation to decision makers. In our view, this is inappropriate, for a number of reasons. More often than not, the different component MOEs will have incommensurate dimensions (e.g., casualties, cost, time) that cannot be combined without using an explicit formula that implicitly weights them. For example, the most common formula is a linear additive weighting scheme. Such a weighting scheme assigns importance (or value) to each of the individual component MOEs, a task that is more appropriately done by the decision maker and not the analyst. Moreover, the many-to-one transformation of the formula may well mask information that is likely to be useful to the decision maker's deliberations.

2. Some MOEs are the ratio of two values, each of which, by itself, is useful in analyzing mission success. However, since both the numerator and the denominator affect the ratio, changes in (or errors in estimating) the numerator have linear effects on the ratio value, while changes (or errors) in the denominator affect the ratio hyperbolically. This effect makes the use of such measures particularly suspect when the denominator can become very small, perhaps even zero. In addition, using a ratio measure to compare a proposed organization or system with an existing one implies a specific value relationship between dimensions of the numerator and the denominator.

Although ratio MOE values may be useful in assessing degrees of mission success, reporting only this ratio may be misleading. Analysis of each of its components will usually be required to interpret the results and develop an understanding of why the mission was successful.

ATEC plans to use IOT data to calculate MOEs and MOPs for the IBCT/Stryker. These data will be collected in two ways: subjectively, using subject-matter experts (SMEs), and objectively, using instrumentation. Our assessment of these plans is presented in the remainder of this chapter, which discusses subjective measures (garnered through the use of SMEs) and objective measures of mission effectiveness and of reliability, availability, and maintainability. SMEs are used to subjectively collect data for MOEs (and MOPs) to assess the performance and effectiveness of a force in both SSC missions (e.g., raid and perimeter defense) and SOSE missions. Objective measures of effectiveness (including casualty-related measures, scenario-specific measures, and system degradation measures) may also be applied across these mission types, although objective casualty-related MOEs are especially useful for evaluating SSC engagements, in which both the IBCT and the OPFOR casualties are indicators of mission success. Casualty-related measures are less commonly applied to SOSE missions, in which enemy losses may have little to do with mission success. Objective measures of reliability, availability, and maintainability are applied to assess the performance and effectiveness of the system.

SUBJECTIVE SUBJECT-MATTER EXPERT MEASURES

Military judgment is an important part of the operational evaluation and will provide the bulk of numerical MOEs for the Stryker IOT. Trained SMEs observe mission tasks and subtasks and grade the results, according to agreed-upon standards and rating scales. The SMEs observe and follow each platoon throughout its mission set. Although two SMEs are assigned to each platoon and make independent assessments, they are not necessarily at the same point at the same time.

SME ratings can be binary (pass/fail, yes/no) judgments, comparisons (e.g., against baseline), or indicators on a numerical task performance rating scale. In addition to assigning a rating, the SME keeps notes with the reasoning behind the assessment. The mix of binary and continuous measures, as well as the fact that the rating scales are not particularly on a cardinal (much less a ratio) scale, makes it inappropriate to combine them in any meaningful way.

Moreover, since close combat tactical training data show that the conventional 10-point rating scale provides values that were rarely (if ever) used by SMEs, ATEC has proposed using an 8-point scale. However, it has also been observed in pretesting that the substantive difference between task performance ratings of 4 and 5 is very much greater than between 3 and 4. This is because, by agreement, ratings between 1 and 4 indicate various levels of task "failure" and ratings between 5 and 8 indicate levels of task "success." The resulting bimodal distribution has been identified by ATEC analysts as representing a technical challenge with respect to traditional statistical analysis. We prefer to regard this phenomenon as being indicative of a more fundamental psychometric issue, having to do with rating scale development and validation. Although there has also been some discussion of using two or three separate rating scales, this would be a useful approach only if there were no attempt to then combine (roll up) these separate scales by means of some arbitrary weighting scheme.

SME judgments are clearly subjective: they combine experience with observations, so that two SMEs could easily come up with different ratings based on the same observations, or a single SME, presented twice with the same observation, could produce different ratings. Using subjective data is by itself no barrier to making sound statistical or operational inferences (National Research Council, 1998b; Veit, 1996). However, to do so, care must be taken to ensure that the SME ratings have the usual properties of subjective data used in other scientific studies, that is, that they can be calibrated, are repeatable, and have been validated. One good way to support the use of SME ratings in an IOT is to present a careful analysis of SME training data, with particular attention paid to demonstrating small inter-SME variance.

OBJECTIVE MEASURES OF EFFECTIVENESS

In this section we discuss objective measures of effectiveness. Although these involve "objective" data, in the sense that two different observers will agree as to their values, experts do apply judgment in selecting the particular variables to be measured in specific test scenarios. While it is useful to provide summary statistics (e.g., for casualty measures, as discussed below), decision makers should also be provided (as we suggest earlier in this chapter) with the values of the component statistics used to calculate summary statistics, since these component statistics may (depending on analytical methods) provide important information in themselves. For example, summary brigade-level casualties (discussed below) are computed by aggregating company and squad-level casualties, which by themselves can be of use in understanding complicated situations, events, and scenarios. There are many thousands of objective component statistics that must support complex analyses that depend on specific test scenarios. Our discussion below of casualty-related measures and of scenario-specific measures is intended to illustrate fruitful analyses.

Casualty-Related Measures

In this section we discuss some of the casualty-related MOEs for evaluating IBCT mission success, appropriate for both combat and SOSE missions, but particularly appropriate for SSC-like engagements in which both sides can inflict significant casualties on each other. Specifically, we discuss the motivation and utility of three casualty ratio MOEs presented by ATEC in its operational test plan.

Ideally, an operational test with unlimited resources would produce estimates of the probability of mission "success" (or any given degree of success), or the distribution of the number of casualties, as a function of force ratios, assets committed and lost, etc. However, given the limited replications of any particular scenario, producing such estimates is infeasible. Still, a variety of casualty-related proxy MOES can be used, as long as they can be shown to correlate (empirically or theoretically) with these ultimate performance measures.

We begin by introducing some notation and conventions.[1]

---

[1] The conventions are based on analyses of the cold war security environment that led to the development and rationale underlying two of the ratio MOEs.

Let:

$N \equiv$ initial number of enemy forces (OPFOR) in an engagement (battle, campaign) against friendly forces
$M \equiv$ initial number of friendly forces in an engagement
$FR_0 \equiv N/M =$ initial force ratio
$n(t) \equiv$ number of surviving enemy forces at time t in the engagement
$m(t) \equiv$ number of surviving friendly forces at time t in the engagement
$FR(t) \equiv n(t)/m(t) \equiv$ force ratio at time t
$C_n(t) = N - n(t) \equiv$ number of enemy casualties by time t
$C_m(t) = M - m(t) \equiv$ number of friendly casualties by time t

Although survivors and casualties vary over time during the engagement, we will drop the time notation for ease in subsequent discussions of casualty-related MOE. In addition, we use the term "casualties" as personnel losses, even though much of the motivation for using ratio measures has been to assess losses of weapon systems (tanks, etc.). It is relatively straightforward to convert a system loss to personnel casualties by knowing the kind of system and type of system kill.

Loss Exchange Ratio[2]

A measure of force imbalance, the loss exchange ratio (LER) is defined to be the ratio of enemy (usually the attacker) losses to friendly (usually defender) losses. That is[3]

$$LER = \frac{N-n}{M-m} = \frac{C_n}{C_m} \qquad (1)$$

---

[2] During the cold war era, measures of warfighting capability were needed to help the Army make resource allocation decisions. The LER measure was created a number of decades ago for use in simulation-based analyses of war between the Soviet-led Warsaw Pact (WP) and the U.S.-led NATO alliance. The WP possessed an overall strategic advantage in armored systems of 2:1 over NATO and a much greater operational-tactical advantage of up to 6:1. Prior to the demise of the Soviet Union in 1989-1991, NATO's warfighting objective was to reduce the conventional force imbalance in campaigns, battles, and engagements to preclude penetration of the Inter-German Border.

[3] Enemy losses will always be counted in the numerator and friendly losses in the denominator regardless of who is attacking.

Thus, LER is an indicator of the degree to which the force imbalance is reduced in an engagement: the force imbalance is clearly being reduced while the condition[4]

$$LER > FR_0 = N/M$$

holds.

Since casualty-producing capability varies throughout a battle, it is often useful to examine the instantaneous *LER*—the ratio of the rates of enemy attacker and defender losses—as a function of battle time t, in order to develop a causal understanding of the battle dynamics. Early in the battle, the instantaneous *LER* is high and relatively independent of the initial force ratio (and particularly threat size) because of concealment and first shot advantages held by the defender. The *LER* advantage moves to the attacker as the forces become decisively engaged, because more attackers find and engage targets, and concentration and saturation phenomena come into play for the attacker.

However, this pattern is not relevant in today's security environment, with new technologies (e.g., precision munitions, second-generation night vision devices, and FBCB2); more U.S. offensive engagements; and threats that employ asymmetric warfare tactics. The utility of the *LER* is further evidenced by its current use by analysts of the TRADOC Analysis Command and the Center for Army Analysis (CAA, formerly the Army's Concepts Analysis Agency) in studies of the Army's Interim Force and Objective Force.

Force Exchange Ratio[5]

The *LER* indicates the degree of mission success in *tactical*-level engagements and allows an examination of the impact of different weapon systems, weapon mixes, tactics, etc. At this level, each alternative in a study traditionally has the same initial U.S. force size (e.g., a battalion, a company). As analysis moves to *operational*-level issues (e.g., force design/struc-

---

[4]The *LER* is usually measured at the time during an engagement when either the attacker or defender reaches a breakpoint level of casualties.

[5]This MOE is also referred to as the fractional loss exchange ratio and the fractional exchange ratio.

ture, operational concepts) with nonlinear battlefields, alternatives in a study often have different initial force sizes. This suggests considering a measure that "normalizes" casualties with respect to initial force size, which gives rise to the force exchange ratio (*FER*):

$$FER = \frac{\left(\dfrac{N-n}{N}\right)}{\left(\dfrac{M-m}{M}\right)} = \frac{\dfrac{C_n}{N}}{\dfrac{C_m}{M}} = \frac{LER}{FR_0} \qquad (2)$$

The *FER* and the *LER* are equally effective as indicators of the degree by which force imbalance is reduced in a campaign: an enemy's initial force size advantage is being reduced as long as the *FER* > 1. Some of the history behind the use of *FER* is summarized in Appendix B.

Relative Loss Ratio

ATEC has proposed using a third casualty ratio, referred to as the relative loss ratio (*RLR*) and, at times, the "odds ratio." They briefly define and demonstrate its computation in a number of documents (e.g., TEMP, December 2001; briefing to USA-OR, June 2002) and (equally briefly) argue for its potential advantages over the *LER* and the *FER*.

The basic *RLR* is defined by ATEC to be the ratio of [enemy to friendly casualty ratio] to [enemy to friendly survivor ratio] at some time t in the battle:

$$RLR = \frac{\dfrac{N-n}{n}}{\dfrac{M-m}{m}} = \frac{\dfrac{C_n}{n}}{\dfrac{C_m}{m}} = \left(\dfrac{C_n}{C_m}\right)\left(\dfrac{m}{n}\right) = LER \cdot SVER \qquad (3)$$

where *SVER* = m/n is referred to as the "survivor ratio." Since the reciprocal of *SVER* is the force ratio $FR_t = (n/m)$ at time t in the battle, *RLR* can be expressed as

$$RLR = \frac{LER}{FR_t} \qquad (4)$$

which is structurally similar to the *FER* given by equation (2). It is interesting to note that the condition

$$RLR = \frac{\frac{N-n}{n}}{\frac{M-m}{m}} > 1$$

implies that

$$FR_0 > FR_t$$

i.e., that the initial force imbalance is being reduced at time t. However, the condition $FER > 1$ also implies the same thing.

ATEC also proposes to use RLR, a relative force ratio normalized for *initial* force ratios. That is:

$$RLR = \frac{\left(\frac{N-n}{N}\right)}{\left(\frac{M-m}{M}\right)} = \frac{\left(\frac{C_n}{N}\right)}{\left(\frac{C_m}{M}\right)} = \frac{\frac{C_n}{N}}{\frac{C_m}{M}} \cdot \frac{m}{n} = FER \cdot SVER = \frac{FER}{FR_t} \quad (5)$$

ATEC does not discuss any specific properties or implications of using the RLR but does suggest a number of advantages of its use relative to the LER and the FER. These are listed below (in italics) with the panel's comments.

1. *The RLR addresses casualties and survivors whereas the LER and the FER address only casualties.* When calculating LER and FER the number of casualties is in fact the initial number of forces minus survivors.

2. *The RLR can aggregate over different levels of force structure (e.g., platoons, companies, battalions) while the LER and the FER cannot.* The initial numbers of forces and casualties for multiple platoon engagements in a company can be aggregated to compute company-level LERs and FERs, and they can be aggregated again over all company engagements to compute battalion-level LERs and FERs. Indeed, this is how they are regularly computed in Army studies of battalion-level engagements.

3. *The RLR can aggregate different kinds of casualties (vehicles, personnel, civilians, fratricide) to present a decision maker with a single RLR measure of merit, while the LER and the FER cannot.* Arbitrary linear additive functions combining these levels of measures are not useful for the reasons given in the section on introduction to measures above. In any event, personnel casualties associated with system/vehicle losses can be readily calculated

using information from the Ballistics Research Laboratories/U.S. Army Material Systems Analysis Activity (BRL/AMSAA). It is not clear why the geometric mean computed for the *RLR* (p. 48 of December 2001 TEMP) could not be computed for the *LER* or the *FER* if such a computation were thought to be useful.

4. *The RLR motivates commanders "to seek an optimum trade-off between friendly survivors and enemy casualties."* This does not appear germane to selecting an MOE that is intended to measure the degree of mission success in the IBCT IOT.

5. *The RLR has numerous attractive statistical properties.* ATEC has not delineated these advantages, and we have not been able to determine what they are.

6. *The RLR has many good statistical properties of a "maximum likelihood statistic" including being most precise among other attractive measures of attrition (LER and FER).* It is not clear what advantage is suggested here. Maximum likelihood estimation is a technique for estimating parameters that has some useful properties, especially with large samples, but maximum likelihood estimation does not appear to address the relative merits of *LER*, *FER*, and *RLR*.

7. *The IAV [Stryker] IOT is a designed experiment. To take advantage of it, there is a standard log-linear modeling approach for analyzing attrition data that uses RLR statistics.* There are equally good statistical approaches that can be used with the *FER* and the *LER*.

Fratricide and Civilian Casualties

ATEC has correctly raised the importance of developing suitable MOEs for fratricide (friendly casualties caused by friendly forces) and civilian casualties caused by friendly fires. It is hypothesized that the IBCT/Stryker weapon capabilities and the capabilities of its C4ISR suite will reduce its potential for fratricide and civilian casualties compared with the baseline. The June 2002 SEP states that in order to test this hypothesis, the "standard" *RLR* and fratricide *RLR* (where casualties caused by friendly forces are used in place of OPFOR casualties) will be compared for both the IBCT and the LIB. A similar comparison would be done using a civilian casualties *RLR*.

However, the *RLR* (as well as the *LER* and the *FER*) is not an appropriate MOE to use, not only for the reasons noted above, but also because it does not consider the appropriate fundamental phenomena that lead to fratricide (or civilian) casualties. These casualties occur when rounds fired at the enemy go astray (for a variety of possible reasons, including erroneous intelligence information, false detections, target location errors, aiming errors, weapons malfunction, etc.). Accordingly, we recommend that ATEC report, as one MOE, the *number* of such casualties for IBCT/Stryker and the baseline force *and also* compute a fratricide frequency (FF) defined as

$$FF = \frac{\text{number of fratricide casualties}}{\text{number of rounds fired at the enemy}}$$

and a similarly defined civilian casualty frequency (CF). The denominator could be replaced by any number of other measures of the intensity (or level) of friendly fire.

Advantages of FER and LER Over RLR

The *FER* and the *LER* have served the Army analysis community well for many decades as mission-level MOEs for campaigns, battles, and engagements. Numerous studies have evidenced their utility and correlation to mission success. Accordingly, until similar studies show that the *RLR* is demonstrably superior in these dimensions, ATEC should use *FER* (and *LER* when appropriate), but not the *RLR*, as the primary mission-level MOE for analyses of engagement results. Our preference for using the *FER*, instead of the *RLR*, is based on the following reasons:

- The *FER* has been historically correlated with the probability of mission success (i.e., winning an engagement/battle), and the *RLR* has not.
- There is strong historical and simulation-based evidence that the *FER* is a valid measure of a force's warfighting capability given its strong correlation with win probability and casualties. It has been useful as a measure of defining "decisive force" for victory.
- The Army analysis community has used, and found useful, *FER* and *LER* as the principal MOEs in thousands of studies involving major theatre war and SSC combat between forces that can inflict noticeable casualties on each other. There is no similar experience with the *RLR*.
- There is no compelling evidence that the purported advantages of the *RLR* presented by ATEC and summarized above are valid. There is little understanding of or support for its properties or value for analysis.
- Using ratio measures such as *FER* and *LER* is already a challenge to the interpretation of results when seeking causal insights. The *RLR* adds

another variable (survivors) to the *LER* ratio (making it more difficult to interpret the results) but does not add any new information, since it is perfectly (albeit negatively) correlated with the casualty variables already included in the *FER* and the *LER*.

Scenario-Specific and System Degradation Measures

ATEC states that the main Army and Department of Defense (DoD) question that needs to be answered during the Stryker operational test is: Is a Stryker-equipped force more effective than the current baseline force? The TEMP states that:

> The Stryker has utility in all operational environments against all projected future threats; however, it is designed and optimized for contingency employment in urban or complex terrain while confronting low- and mid-range threats that may display both conventional and asymmetric warfare capabilities.

This statement points directly to the factors that have been used in the current test design: terrain (rural and urban), OPFOR intensity (low, medium, high), and mission type (raid, perimeter defense, security operations in a stability environment). These factors are the ones ATEC wants to use to characterize *if and when* the Stryker-equipped force is better than the baseline and to help explain *why*.

The Stryker SEP defines effectiveness and performance criteria and assigns a numbering scheme to these criteria and their associated measures. In the discussion below, the numbering of criteria adheres to the Stryker SEP format (U.S. Department of Defense, 2002c). There are three sets of measures that are appropriate for assessing each of the three mission types. These are detailed in the measures associated with Criterion 4-1: Stryker systems must successfully support the accomplishment of required operations and missions based on standards of performance matrices and associated mobility and performance requirements.

In particular, the measures of effectiveness for Criterion 4-1 are:

MOE 4-1-1   Mission accomplishment.
MOE 4-1-2   Performance ratings on selected tasks and subtasks from the applicable performance assessment matrices while conducting operations at company, platoon, squad, and section level.
MOE 4-1-3   Relative attrition.

These measures of effectiveness have been addressed in the previous sections.

In addition, however, ATEC would like to know *why* there are differences in performance between the Stryker-equipped force and the baseline force. The reasons for performance differences can be divided into two categories: Stryker capabilities and test factors.

Stryker capabilities include situation awareness (which contributes to survival by avoidance), responsiveness, maneuverability, reliability-availability-maintainability (RAM), lethality, survivability (both ballistic and nonballistic), deployability, transportability, and logistics supportability. Test factors include time of day, time of year, weather, nuclear/biological/chemical (NBC) environment, personnel, and training. Measures for reliability are addressed in detail later in this chapter; test factors are addressed in Chapter 4.

With the exception of situation awareness, responsiveness, maneuverability, and RAM, the current SEP addresses each capability using more of a technical than an operational assessment. The IBCT/Stryker IOT is not designed to address (and cannot be redesigned to address) differences in performance due to lethality, survivability, deployability, transportability, or logistics supportability. Any difference in performance that might be attributed to these factors can only be assessed using the military judgment of the evaluator supported by technical and developmental testing and modeling and simulation.

The current capability measures for situation awareness, responsiveness, and maneuverability are associated with Criterion 4-2 (the Stryker systems must be capable of surviving by avoidance of contact through integration of system speed, maneuverability, protection, and situation awareness during the conduct of operations) and Criterion 4-3 (the Stryker must be capable of hosting and effectively integrating existing and planned Army command, control, communications, computers, intelligence, surveillance, and reconnaissance or C4ISR systems).

The associated MOEs are:

MOE 4-2-1   Improvement of force protection
MOE 4-2-2   Improvement in mission success attributed to information
MOE 4-2-3   Contributions of Army battle command systems (ABCS) information to Stryker survival

MOE 4-2-4 How well did the ABCS allow the commander and staff to gain and maintain situation awareness/understanding?
MOE 4-3-1 Ability to host C4ISR equipment and its components
MOE 4-3-2 Integration effectiveness of C4ISR demonstrated during the product verification test
MOE 4-3-3 Interoperability performance for the Stryker C4ISR in technical testing
MOE 4-3-4 Capability of the Stryker C4ISR to withstand external and internal environmental effects IAW MIL-STD 810F and/or DTC Test Operation Procedures (TOP)
MOE 4-3-5 Capability to integrate MEP and FBCB2 data

The measures associated with Criterion 4-3 are primarily technical and address the ability of the existing hardware to be integrated onto the Stryker platforms. As with many of the other capabilities, any difference in performance that might be attributed to hardware integration will be assessed using the military judgment of the evaluator supported by technical and developmental testing.

The problem with most of the MOPs associated with Criterion 4-2 (see Table 3-1) is that they are not unambiguously measurable. For example, consider MOP 4-2-2-2, communications success. The definition of success is, of course, very subjective, even with the most rigorous and validated SME training. Moreover, the distinction between transfer of information and the value of the information is important: communications can be successful in that there is a timely and complete transfer of critical information, but at the same time unsuccessful if that information is irrelevant or misleading. Or, for another example, consider: MOP 4-2-1-3, Incidents of BLUFOR successful avoidance of the adversary. Whether this criterion has been met can be answered only by anecdote, which is not usually considered a reliable source of data. Note that there is no clear numerator or denominator for this measure, and merely counting the frequency of incidents does not provide a reference point for assessment.

Two other categories of measures that could be more useful in assessing performance differences attributable to situation awareness, responsiveness, and maneuverability are scenario-specific and system degradation measures.

TABLE 3-1 MOPs for Criterion 4-2

MOE 4-2-1	Improvement in force protection
MOP 4-2-1-1	Relative attrition
MOP 4-2-1-2	Mission success rating
MOP 4-2-1-3	Incidents of BLUFOR successful avoidance of the adversary
MOP 4-2-1-4	Incidents where OPFOR surprises the BLUFOR
MOE 4-2-2	Improvement in mission success attributed to information
MOP 4-2-2-1	Initial mission, commander's intent and concept of the operations contained in the battalion and company operations and fragmentary orders
MOP 4-2-2-2	Communications success (use MOE 4-3-5: Capability to integrate MEP and FBCB2 data)
MOE 4-2-3	Contributions of ABCS information (C2, situation awareness, etc.) to Stryker survival
MOP 4-2-3-1	What were the ABCS message/data transfer completion rates (MCR)?
MOP 4-2-3-2	What were the ABCS message/data transfer completion times (speed of service)?
MOP 4-2-3-3	How timely and relevant/useful was the battlefield information (C2 message, targeting information, friendly and enemy situation awareness updates, dissemination of order and plans, alerts and warning) provided by ABCS to commander and staffs?
MOE 4-2-4	How well did the ABCS allow the commander and staff to gain and maintain situation awareness/understanding?
MOP 4-2-4-1	Friendly force visibility
MOP 4-2-4-2	Friendly position data distribution
MOP 4-2-4-3	Survivability/entity data distribution

Scenario-Specific Measures

Scenario-specific measures are those that are tailored to the exigencies of the particular mission-script combinations used in the test. For example, in the perimeter defense mission, alternative measures could include answers to questions such as:

- Did the red force penetrate the perimeter? How many times?
- To what extent was the perimeter compromised (e.g., percentage of perimeter compromised, taking into account the perimeter shape)?
- How far in from the perimeter was the red force when the penetration was discovered?

- How long did it take the red force to penetrate the perimeter?
- What fraction of time was the force protected while the OPFOR was (or was not) actively engaged in attacking the perimeter?

For a raid (or assault) mission, measures might include:

- Was the objective achieved?
- How long did it take to move to the objective?
- How long did it take to secure the objective?
- How long was the objective held (if required)?

For SOSE missions, measures might include:

- For "show the flag" and convoy escort: How far did the convoy progress? How long did it take to reach the convoy? How much time transpired before losses occurred?
- For route and reconnaissance: How much information was acquired? What was the quality of the information? How long did it take to acquire the information?

We present here the principle that useful objective measures can be tied to the specific events, tasks, and objectives of missions (the unit of measurement need not always be at the mission level or at the level of the individual soldier), and so the measures suggested are intended as exemplary, not as exhaustive. Other measures could easily be tailored to such tasks as conducting presence patrols, reaching checkpoints, searching buildings, securing buildings, enforcing curfews, etc. These kinds of measures readily allow for direct comparison to the baseline, and definitions can be written so that they are measurable.

System Degradation Measures: Situation Awareness as an Experimental Factor

The other type of measure that would be useful in attributing differences to a specific capability results from degrading this capability in a controlled manner. The most extreme form of degradation is, of course, complete removal of the capability. One obvious Stryker capability to test in this way is situation awareness. The IBCT equipped with Stryker is intended to provide more combat effectiveness than the LIB and be more strategically deployable than a heavy MIB. More combat effectiveness is achieved by providing the IBCT with significantly more firepower and tactical mobility (vehicles) than the LIB. Improving strategic mobility is provided by designing the IBCT systems with significantly less armor, thus making them lighter than systems in the heavy MIB, but at a potential price of being more vulnerable to enemy fire. The Army has hypothesized that this potential vulnerability will be mitigated by Striker's significantly improved day and night situation awareness and C4ISR systems such as FBCB2,[6] second-generation forward-looking infrared systems, unmanned aerial vehicles, and other assets.

If all C4ISR systems perform as expected and provide near-perfect situation awareness, the IBCT should have the following types of advantages in tactical engagements over the LIB (which is expected to have much less situation awareness):

- IBCT units should be able to move better (faster, more directly) by taking advantage of the terrain and having common knowledge of friendly and enemy forces.
- With better knowledge of the enemy, IBCT units should be able to get in better positions for attack engagements and to attack more advantageously day or night by making effective use of cover in approaches to avoid enemy fires. They could structure attacks against the enemy in two directions (thus making him fight in two directions) with little or no risk of surprise ambushes by threat forces.
- IBCT units and systems should be able to acquire more enemy targets accurately at longer ranges, especially at night, facilitating more effective long-range fire.
- IBCT systems should be able to rapidly "hand off" targets to enhance unit kill rates at all ranges.
- Using combinations of the above situation awareness advantages, IBCT units should be capable of changing traditional attacker-defender battle dynamics favoring the defender at long ranges and the attacker at shorter ranges. Attacking IBCT systems should be able to avoid long-range defender fires or attrit many of the defenders at long range before closing with them.

---

[6]FBCB2 is a top-down fed command and control system that is supposed to provide the IBCT with timely and accurate information regarding all friendly and enemy systems.

The Army has yet to test the underlying hypothesis that the enhanced situation awareness/C4ISR will in fact make the IBCT/Stryker less vulnerable and more effective. As currently designed, the IOT (which compares the effectiveness of IBCT/Stryker with the LIB in various missions) cannot test this hypothesis since the IBCT/Stryker is presumably more effective than the LIB for many criteria (mobility, lethality, survivability, etc.), not just in its situation awareness/C4ISR capability. To most effectively test the underlying hypothesis, the IOT design should make situation awareness/C4ISR an explicit factor in the experiment, preferably with multiple levels, but at a minimum using a binary comparison. That is, the design should be modified to explicitly incorporate trials of the IBCT/Stryker both with and without its improved situation awareness/C4ISR in both daytime and nighttime scenarios.

It is not sufficient to rely on test conditions (e.g., the unreliability of the hardware itself) to provide opportunities to observe missions without situation awareness. There must be a scripted turning off of the situation awareness hardware. This kind of controlled test condition leads to results that can be directly attributed to the situation awareness capability.

If this type of test modification is not feasible, then the underlying hypothesis should be tested using appropriate simulations at either the Intelligence School or TRAC-FLVN (Ft. Leavenworth). Although the hypothesis may not be testable in the IOT as currently designed, ATEC may be able to determine some of the value of good situation awareness/C4ISR by assessing the degree to which the situation awareness-related advantages noted above are achieved by the IBCT/IAV in combat missions. To accomplish this:

- SMEs should assess whether the IBCT/Stryker units move through the terrain better (because of better information, not better mobility) then LIB units.
- SMEs should assess whether IBCT/Stryker units get in better positions (relative to enemy locations) for attack engagements than LIB units and are able to design and implement attack plans with more covered attack routes to avoid enemy fires (i.e., reduce their vulnerability).
- ATEC should collect target acquisition data by range and by type (visual, pinpoint) for day and night missions to determine whether IBCT/Stryker systems have the potential for more long-range fires than LIB systems. ATEC should also record the time and range distribution of actual fire during missions.

- ATEC should determine the number of hand-off targets during engagements to see if the IBCT force is really more "net-centric" than the LIB.
- From a broader perspective, ATEC should compute the instantaneous *LER* throughout engagements to see if improved situation awareness/C4ISR allows the IBCT force to advantageously change traditional attacker-defender battle dynamics.

OBJECTIVE MEASURES OF SUITABILITY

The overall goal of the IOT is to assess baseline force versus IBCT/Stryker force effectiveness. Because inadequate levels of reliability and maintainability (R&M) would degrade or limit force effectiveness, R&M performance is important in evaluating the Stryker system. We note in passing that R&M performance will affect both sides of the comparison. It is not clear whether an assessment of baseline R&M performance is envisioned in the IOT. Such an assessment would provide an important basis for comparison and might give insights on many differences in R&M effectiveness.

Reliability

Criterion 1-3 states: "The Stryker family of interim armored vehicles (excluding GFE components and systems) will have a reliability of 1,000 mean miles between critical failures (i.e., system aborts)." This requirement is raised to 2,000 mean miles for some less stressed vehicle types. These failures could be mechanical vehicle failures or failures due to vehicle/GFE interface issues. Although GFE failures themselves don't contribute to this measure, they should and will be tracked to assess their role in the force effectiveness comparison.

The IOT is not only key to decisions about meeting R&M criteria and systems comparisons, but it also should be viewed as a shakedown exercise. The IOT will provide the first view of the many mechanical and electronic pieces of equipment that can fail or go wrong in an operational environment. Some failures may repeat, while others will take a fair amount of IOT exposure to manifest themselves for the first time. Thus the IOT provides an opportunity for finding out how likely it is that other new failure issues may crop up.

For this reason, failure incidents should be collected for all vehicles for their entire lives on a vehicle-by-vehicle basis, even though much of the data may not serve the express purposes of the IOT. Currently it appears that only the Army test incident reporting system will be used. Suitable databases to maintain this information should be established.

In the remainder of this section we discuss four important aspects of reliability and maintainability assessment:

- failure modes (distinguishing between them and modeling their failure time characteristics separately);
- infant mortality, durability/wearout, and random failures (types and consequences of these three types of failure modes);
- durability—accelerated testing and add-on armor; and
- random failures, GFE integration, and scoring criteria.

Failure Modes

Although the TEMP calls for reporting the number of identified failures and the number of distinct failure modes, these are not sufficient metrics for making assessments about systems' R&M. Failures need to be classified by failure mode. Those modes that are due to wearout have different data-recording requirements from those that are due to random causes or infant mortality. For wearout modes, the life lengths of the failed parts/systems should be observed, as well as the life lengths of all other equivalent parts that have not yet failed. Life lengths should be measured in the appropriate time scale (units of operating time, or operating miles, whichever is more relevant mechanistically). Failure times should be recorded *both* in terms of the life of the vehicle (time/miles) and in terms of time since last maintenance. If there are several instances of failure of the same part on a given vehicle, a record of this should be made. If, for example, the brake or tire that fails or wears out is always in the same position, this would be a significant finding that would serve as input for corrective action.

Different kinds of failure modes have different underlying hazard functions (e.g., constant, increasing, or decreasing). When considering the effect of R&M on system effectiveness, it is potentially misleading to report the reliability of a system or subsystem in terms of a MOP that is based on a particular but untested assumption. For example, reporting of only the "mean time to failure" is sufficiently informative only when the underlying failure time distribution has only a single unknown parameter, such as a constant hazard function (e.g., an exponential distribution). One alternative is to report reliability MOPs separately for random types of failure modes (constant hazard function), wearout failure modes (increasing hazard function), and defect-related failure modes (decreasing hazard function). These MOPs can then be used to assess the critical reliability performance measure: the overall probability of vehicle failure during a particular future mission.

Wearout failures may well be underrepresented in the IOT, since most vehicles are relatively new. They also depend heavily on the age mix of the vehicles in the fleet. For that reason, and to correct for this underrepresentation, it is important to model wearout failures separately.

Some measure of criticality (not just "critical" or "not critical") should be assigned to each failure mode so as to better assess the effect(s) of that mode. Further subdivision (e.g. GFE versus non-GFE) may also be warranted.

Data on the arrival process of new failure modes should be carefully documented, so that they can be used in developing a model of when new failure modes occur as a function of fleet exposure time or miles. The presumably widening intervals[7] between the occurrence of new failure modes will enable an assessment of the chance of encountering any further and as yet unseen failure modes. The use of these data to make projections about the remaining number of unseen failure modes should be done with great care and appreciation of the underlying assumptions used in the projection methodology.

Although the different Stryker vehicle variants will probably have different failure modes, there is a reasonable possibility that information across these modes can be combined when assessing the reliability of the family of vehicles.

In the current TEMP, failure modes from developmental test (DT) and IOT are to be assessed across the variants and configurations to determine the impact that the operational mission summary/mission profile and unique vehicle characteristics have on reliability estimates. This assessment can be handled by treating vehicle variant as a covariate. Other uncontrollable covariates, such as weather conditions, could certainly have an impact, but it is not clear whether these effects can be sorted out cleanly. For

---

[7]Of course, these widening intervals are not likely to be true in the immediate period of transferring from developmental test to operational test, given the distinct nature of these test activities.

example, one could record the degree of wetness of soil conditions on a daily basis. This might help in sorting out the potential confounding of weather conditions under which a given force (IBCT or baseline) is operating. For example, if the IBCT were to run into foul weather halfway through its testing, and if certain failures appeared only at that time, one would be able to make a better case for ascribing the failures to weather rather than to the difference in force, especially if the baseline force does not run into foul weather.

Infant Mortality

Operational tests, to some extent, serve the purpose of helping to uncover and identify unknown system design flaws and manufacturing problems and defects. Such "infant mortality" problems are normally corrected by making design or manufacturing changes or through the use of sufficient burn-in so that the discovered infant mortality failure modes will no longer be present in the mature system.

The SEP describes no specific MOPs for this type of reliability problem. Indeed, the SEP R&M MOPs (e.g., estimates of exponential distribution mean times) assume a steady-state operation. Separate measures of the effects of infant mortality failures and the ability to eliminate these failure modes would be useful for the evaluation of Stryker system effectiveness.

Durability and Wearout

The IOT currently has no durability requirement, but issues may come up in the evaluation. Vehicles used in the IOT will not have sufficient operating time to produce reliable R&M data in general and especially for durability. Although the SEP mentions an historical 20,000-mile durability requirement, the Stryker system itself does not have a specified durability requirement. ATEC technical testing will, however, look at durability of the high-cost components. In particular, in DT, the infantry carrier vehicle will be tested in duration tests to 20,000 miles.

Add-On Armor

Whether or not vehicles are outfitted with their add-on armor (AoA) can be expected to have an important impact on certain reliability metrics. The AoA package is expected to increase vehicle weight by 20 percent. The added weight will put additional strain on many operating components, particularly the vehicle power train and related bearings and hydraulic systems. The additional weight can be expected to increase the failure rate for all types of failure modes: infant mortality, random, and, especially, durability/wear. Because product verification test (PVT) and DT will be done in understressed conditions (that is, without AoA), any long-term durability problems that do show up can be judged to be extremely serious, and other problems that may exist are unlikely to be detected in IOT. Although the IOT will proceed without AoA (because it will not be ready for the test), weight packs should be used even if there is currently imperfect knowledge about the final weight distribution of the AoA. Doing this with different weight packs will go a long way to assess the impact of the weight on the reliability metrics. The details of the actual AoA weight distribution will presumably amount to only a small effect compared with the effect of the presence or absence of armor.

There is a need to use PVT, DT, and IOT results to support an early fielding decision for Stryker. Because of the absence of valid long-term durability data under realistic operating conditions (i.e., with AoA installed), the planned tests will not provide a reasonable degree of assurance that Stryker will have durability that is sufficient to demonstrate long-term system effectiveness, given the potential for in-service failure of critical components.

Some wearout failure modes (not necessarily weight-related) may show up during the IOT, but they are likely to be underrepresented compared with steady-state operation of the Stryker fleet, because the vehicles used in the IOT will be relatively new. For such failure modes it is important to capture the time to failure for each failed part/system and the time exposed without failure for each other equivalent part/system. This will enable correction for the underreporting of such failure modes and could lead to design or maintenance changes.

Random Failures, GFE, and Scoring Criteria

Random failures are those failures that are not characterized as either infant mortality or durability/wearout failures. These should be tracked by vehicle type and failure mode. Random failures are generally caused by events external to the system itself (e.g., shocks or accidents). The excessive occurrence of random failures of a particular failure mode during IOT may indicate the need for system design changes to make one or more vehicle types more robust to such failure modes. Because of such potential experiences, it is important to track all of these random failure modes separately, even though it is tempting to lump them together to reduce paperwork requirements.

The reliability of the GFE integration is of special concern. The blending of GFE with the new physical platform may introduce new failure modes at the interface, or it may introduce new failure modes for the GFE itself due to the rougher handling and environment. R&M data will be analyzed to determine the impact of GFE reliability on the system and the GFE interfaces. Although GFE reliability is not an issue to be studied by itself in IOT, it may have an impact on force effectiveness, and for this reason R&M GFE data should be tracked and analyzed separately. Since the GFE on Stryker is a software-intensive system, software failure modes can be expected to occur. To the extent possible, MOPs that distinguish among software-induced failures in the GFE, other problems with the GFE, and failures outside the GFE need to be used.

R&M test data (e.g., test incidents) will be evaluated and scored at an official R&M scoring conference in accordance with the Stryker failure definition/scoring criteria. R&M MOPs will be calculated from the resulting scores. Determination of mission-critical failure modes should not, however, be a binary decision. Scoring should be on an interval scale between 0 and 1 rather than being restricted to 0 (failure) or 1 (nonfailure). For example, reporting 10 scores of 0.6 and 10 scores of 0.4 sends a different message, and contains much more information, than reporting 10 scores of 1 and 10 scores of 0.

We also suggest the use of standard language in recording events to make scoring the events easier and more consistent. The use of standard language also allows for combining textual information across events and analyzing the failure event database.

Availability and Maintainability

MOPs for availability/maintainability, described in the SEP, include mean time to repair; the chargeable maintenance ratio (the ratio of chargeable maintenance time to the total amount of operating time); and preventive maintenance, checks, and services time required. Although these MOPs will be evaluated primarily using data obtained during DT, IOT information should be collected and used to complement this information.

Given that some reliability criteria are expressed as number of failures per 1,000 miles, and since repair time is not measured in miles, an attempt should be made to correlate time (operating time, mission time) with miles so that a supportable comparison or translation can take place.

Contractors do initial maintenance and repair and then train the soldiers to handle these tasks. MOPs computed on the basis of DT-developed contract maintainers and repairmen may not accurately reflect maintainability and repair when soldiers carry out these duties. Therefore, contractor and soldier maintenance and repair data should not be pooled until it has been established that repair time distributions are sufficiently close to one another.

SUMMARY

Reporting Values of Measures of Effectiveness

1. Different MOEs should not be rolled up into a single overall number that tries to capture effectiveness or suitability.

2. Although ratio MOE values may be useful in assessing degrees of mission success, both the numerator and the denominator should be reported.

Subject-Matter Expert Measures

3. To help in the calibration of SME measures, each should be asked to review his or her own assessment of the Stryker IOT missions, for each scenario, immediately before he or she assesses the baseline missions (or vice versa).

4. ATEC should review the opportunities and possibilities for SMEs to contribute to the collection of objective data, such as times to complete certain subtasks, distances at critical times, etc.

5. The inter-SME rating variances from training data should be considered to be the equivalent of instrument error when making statistical inferences using ratings obtained from IOT.

6. The correlation between SME results and objective measures should be reported for each mission.

7. ATEC should consider using two separate SME rating scales: one for "failures" and another for "successes."

8. As an alternative to the preceding recommendation, SMEs could assign ratings on a qualitative scale (for example, the five-point scale: "excel-

lent," "good," "fair," "poor," and "unsatisfactory"). Any subsequent statistical analysis, particularly involving comparisons, would then involve the use of techniques suitable for ordered categorical variables.

9. If resources are available, more than one SME should be assigned to each unit and trained to make independent evaluations of the same tasks and subtasks.

Objective Casualty-Related Measures

10. *FER* (and the *LER* when appropriate), but not the *RLR*, should be used as the primary mission-level MOE for analyses of engagement results.

11. ATEC should use fratricide frequency and civilian casualty frequency (as defined in this chapter) to measure the amount of fratricide and collateral damage in a mission.

Objective Scenario-Specific and System Degradation Measures

12. Only MOPs that are unambiguously measurable should be used.

13. Scenario-specific MOPs should be added for SOSE missions.

14. Situation awareness should be introduced as an explicit test condition.

15. If situation awareness cannot be added as an explicit test condition, additional MOPs (discussed in this chapter) should be added as indirect measures of situation awareness.

16. ATEC should use the "instantaneous *LER*" measure to determine changes in traditional attacker/defender engagement dynamics due to improved situation awareness.

Measures of Reliability and Maintainability

17. The IOT should be viewed as a shakedown process and an opportunity to learn as much as possible about the RAM of the Stryker.

18. RAM data collection should be an ongoing enterprise. Failure and maintenance information should be tracked on a vehicle or part/system basis for the entire life of the vehicle or part/system. Appropriate databases should be set up. This was probably not done with those Stryker vehicles already in existence but it could be implemented for future maintenance actions on all Stryker vehicles.

19. With respect to the difficulty of reaching a decision regarding reli-

ability, given limited miles and absence of add-on armor, weight packs should be used to provide information about the impact of additional weight on reliability.

20. Accelerated testing of specific system components prior to operational testing should be considered in future contracts to enable testing in shorter and more realistic time frames.

21. Failure modes should be considered separately rather than trying to develop failure rates for the entire vehicle using simple exponential models. The data reporting requirements vary depending on the failure rate function.

4

Statistical Design

In this chapter we first discuss some broader perspectives and statistical issues associated with the design of any large-scale industrial experiment. We discuss the designs and design processes that could be implemented if a number of constraints in the operational test designs were either relaxed or abandoned. Since the operational test design for the IBCT/Stryker is now relatively fixed, the discussion is intended to demonstrate to ATEC the advantages of various alternative approaches to operational test design that could be adopted in the future, and therefore the need to reconsider these constraints. This is followed by a brief description of the current design of the IBCT/Stryker initial operational test (IOT), accompanied by a review of the design conditioned on adherence to the above-mentioned constraints.

BROAD PERSPECTIVE ON EXPERIMENTAL DESIGN OF OPERATIONAL TESTS

Constrained Designs of ATEC Operational Tests

ATEC has designed the IOT to be consistent with the following constraints:

1. Aside from statistical power calculations, little information on the performance of IBCT/Stryker or the baseline Light Infantry Brigade

(LIB)—from modeling or simulation, developmental testing, or the performance of similar systems—is used to impact the allocation of test samples in the test design. In particular, this precludes increasing the test sample size for environments for which the IBCT/Stryker or the baseline has proved to be problematic in previous tests.

2. The allocation of test samples to environments is constrained to reflect the allocation of use detailed in the operational mission summary/mission profile (OMS/MP).

3. Operational tests are designed to test the system for typical stresses that will be encountered in the field. This precludes testing systems in more extreme environments to provide information on the limitations of system performance.

4. Possibly most important, operational tests are, very roughly speaking, single test events. It is currently not typical for an operational test either to be carried out in stages or to include use of smaller-scale tests with operationally relevant features focused on specific issues of interest.

Reconsidering Operational Test Design: Initial Operational Testing Should Not Commence Until System Design Is Mature

The above constraints do not need strict adherence, which will result in designs that have substantial disadvantages compared with current methods used in industrial settings. The following discussion provides some characteristics of operational test designs that could be implemented if these constraints were relaxed or removed.

There are two broad goals of any operational test: to learn about a system's performance and its performance limitations in a variety of settings and to confirm either that a system meets its requirements or that it outperforms a baseline system (when this is with respect to average performance over a variety of environments). A fundamental concern with the current approach adopted by ATEC is that both of these objectives are unlikely to be well addressed by the same test design and, as a result, ATEC has (understandably) focused on the confirmatory objective, with emphasis on designs that support significance testing.

Given either a learning or a confirmatory objective, a requisite for operational testing is that it should not commence until the system design is mature. Developmental testing should be used to find major design flaws, including many of those that would typically arise only in operationally realistic conditions. Even fine-tuning the system to improve performance should be carried out during developmental testing. This is especially true for suitability measures. Operational testing performs a difficult and crucial role in that it is the only test of the system as a whole in realistic operational conditions. Operational testing can be used to determine the limitations and value, relative to a baseline system, of a new system in realistic operational conditions in carrying out various types of missions. While operational testing can reveal problems that cannot be discovered, or discovered as easily, in other types of testing, the primary learning that should take place during operational test should be the development of a better understanding of system limitations, i.e., the circumstances under which the system performs less well and under which the system excels (relative to a baseline system). Discovering major design flaws during an operational test that could have been discovered earlier compromises the ability of the operational test to carry out these important functions.

The benefits of waiting until a system design is mature before beginning operational testing does not argue against the use of spiral development. In that situation, for a given stage of acquisition, one should wait until that stage of development is mature before entering operational test. That does not then preclude the use of evolutionary acquisition for subsequent stages of development. (This issue is touched on in Chapter 6.)

Multiple Objectives of Operational Testing and Operational Test Design: Moving Beyond Statistical Significance as a Goal

Operational test designs need to satisfy a number of objectives. Major defense systems are enormously complicated, with performances that can change in important ways as a result of changes in many factors of interest. Furthermore, there are typically dozens of measures for which information on performance is needed. These measures usually come in two major types — those used to compare a new system with a baseline system[1] and those used to compare a new system with its requirements, as provided in the Operational Requirements Document (ORD). In nearly all cases, it is impossible to identify a single operational test design that is simultaneously best for identifying how various factors affect system performance for doz-

---

[1] Though not generally feasible, the use of multiple baselines should sometimes be considered, since for some environments some baselines would provide little information as comparison systems.

ens of measures of interest. Test designs that would be optimal for the task of comparing a system with requirements would not generally be as effective for comparing a system with a baseline, and test designs that would be optimal for measures of suitability would not generally be excellent for measures of effectiveness. In practice, one commonly selects a primary measure (one that is of greatest interest), and the design is selected to perform well for that measure. The hope is that the other measures of interest will be related in some fashion to the primary measure, and therefore the test design to evaluate the primary measure will be reasonably effective in evaluating most of the remaining measures of interest. (If there are two measures of greatest interest, a design can be found that strikes a balance between the performance for the two measures.)

In addition, operational tests can have a number of broader goals:

1. to understand not only how much the various measures differ for the two systems but also why the measures differ;

2. to identify additional unknown factors that affect system performance or that affect the difference between the operation of the system being tested and the baseline system;

3. to acquire a better strategic understanding of the system, for example, to develop a greater understanding of the value of information, mobility, and lethality for performance;

4. to understand the demands on training and the need for system expertise in operating the system in the field; and

5. to collect sufficient information to support models and simulations on system performance.

The test and evaluation master plan (TEMP) states that the Stryker:

> has utility in all operational environments against all projected future threats; however, it is designed and optimized for contingency employment in urban or complex terrain while confronting low- and mid-range threats that may display both conventional and asymmetric warfare capabilities.

Clearly, the operational test for Stryker will be relied on for a number of widely varied purposes.

As stated above, ATEC's general approach to this very challenging problem focuses on the objective of confirming performance and uses the statistical concept of significance testing: comparing the performance of IBCT/Stryker against the baseline (LIB) to establish that the former is preferred to the latter. In addition, there is some testing against specific requirements (e.g., Stryker has a requirement for 1,000 mean miles traveled between failures). This approach, which results in the balanced design described in Chapter 2 (for a selected number of test design factors), does not provide as much information as other approaches could in assessing the performance of the system over a wide variety of settings.

To indicate what might be done differently in the IBCT/Stryker IOT (and for other systems in the future), we discuss here modifications to the sample size, test design, and test factor levels.

Sample Size

Given that operational tests have multiple goals (i.e., learning and confirming for multiple measures of interest), arguments for appropriate sample sizes for operational tests are complicated. Certainly, sample sizes that support minimal power at reasonable significance levels for testing primary measures of importance provide a starting point for sample size discussions. However, for complicated, expensive systems, given the dynamic nature of system performance as a function of a number of different factors of importance (e.g., environments, mission types), it is rare that one will have sufficient sample size to be able to achieve adequate power. (Some benefits in decreasing test sample size for confirmatory purposes can be achieved through use of sequential testing, when feasible.) Therefore, budgetary limitations will generally drive sample size calculations for operational tests. However, when that is not the case, the objectives of learning about system performance, in addition to that of confirming improvement over a baseline, argue for additional sample size so that these additional objectives can be addressed. Therefore, rather than base sample size arguments solely on power calculations, the Army needs to allocate as much funding as various external constraints permit to support operational test design.

Testing in Scenarios in Which Performance Differences Are Anticipated

As mentioned above, ATEC believes that it is constrained to allocate test samples to mission types and environments to reflect expected field use, as provided in the OMS/MP. This constraint is unnecessary, and it works against the more important goal of understanding the differences between the IBCT/Stryker and the baseline and the causes of these differences. If a specific average (one that reflects the OMS/MP) of performance across mission type is desired as part of the test evaluation, a reweighting of the estimated performance measures within scenario can provide the desired summary measures a posteriori. Therefore, the issue of providing specific averages in the evaluation needs to be separated from allocation of test samples to scenarios.

As indicated above, test designs go hand-in-hand with test goals. If the primary goal for ATEC in carrying out an operational test is to confirm that, for a specific average over scenarios that conforms to the OMS/MP missions and environments, the new system significantly outperforms the baseline, then allocations that mimic the OMS/MP may be effective. However, if the goal is one of learning about system performance for each scenario, then, assuming equal variances of the performance measure across scenarios, the allocation of test samples equally to test scenarios would be preferable to allocations that mimic the OMS/MP.

More broadly, general objectives for operational test design could include: (1) testing the average performance across scenarios (reflecting the OMS/MP) of a new system against its requirements, (2) testing the average performance of a new system against the baseline, (3) testing performance of a new system against requirements or against a baseline for individual scenarios, or (4) understanding the types of scenarios in which the new system will outperform the baseline system, and by how much. Each of these goals would generally produce a different optimal test design.

In addition to test objectives, test designs are optimized using previous information on system performance, which are typically means and variances of performance measures for the system under test and for the baseline system. This is a catch-22 in that the better one is able to target the design based on estimates of these quantities, the less one would clearly need to test, because the results would be known. Nevertheless, previous information can be extremely helpful in designing an operational test to allocate test samples to scenarios to address test objectives.

Specifically, if the goal is to obtain high power, within each scenario, for comparing the new system with the baseline system on an important measure, then a scenario in which the previous knowledge was that the mean performance of the new system was close to that for the baseline would result in a large sample allocation to that scenario to identify which system is, in fact, superior. But if the goal is to better understand system performance within the scenarios for which the new system outperforms the baseline system, previous knowledge that the mean performances were close would argue that test samples should be allocated to other test scenarios in which the new system might have a clear advantage.

Information from developmental tests, modeling and simulation, and the performance of similar systems with similar components should be used to target the test design to help it meet its objectives. For IBCT/Stryker, background documents have indicated that the Army expects that differences at low combat intensity may not be practically important but that IBCT/Stryker will be clearly better than the baseline for urban and high-intensity missions. If the goal is to best understand the performance of IBCT/Stryker in scenarios in which it is expected to perform well, it would be sensible to test very little in low-intensity scenarios, since there are unlikely to be any practical and statistically detectable differences in the performance between IBCT/Stryker and the baseline. Understanding the advantages of IBCT/Stryker is a key part of the decision whether to proceed to full-rate procurement; therefore, understanding the degree to which Stryker is better in urban, high-intensity environments is important, and so relatively more samples should be allocated to those situations. There may be other expectations concerning IBCT/Stryker that ATEC could comfortably rely on to adapt the design to achieve various goals.

Furthermore, because the baseline has been used for a considerable length of time, its performance characteristics are better understood than those of IBCT/Stryker. While this may be less clear for the specific scenarios under which IBCT/Stryker is being tested, allocating 42 scenarios to the baseline system may be inefficient compared with the allocation of greater test samples to IBCT/Stryker scenarios.

Testing with Factors at High Stress Levels

A general rule of thumb in test design is that testing at extremes is often more informative than testing at intermediate levels, because information from the extremes can often be used to estimate what would have happened at intermediate levels. In light of this, it is unclear how extreme the high-intensity conflict is, as currently scripted. For example, would the use of 300 OPFOR players be more informative than current levels? Our impression is that, in general, operational testing tests systems at typical stress levels. If testing were carried out in somewhat more stressful situations than are likely to occur, information is obtained about when a system is likely to start breaking down, as well as on system performance for typical levels of stress (although interpolation from the stressful conditions back to typical conditions may be problematic). Such a trade-off should be considered in the operational test design for IBCT/Stryker. In the following section, a framework is suggested in which the operational test is separated into a learning component and a confirming component. Clearly, testing with factors at high stress levels naturally fits into the learning component of that framework, since it is an important element in developing a complete understanding of the system's capabilities and limitations.

Alternatives to One Large Operational Test

In the National Research Council's 1998 report *Statistics, Testing, and Defense Acquisition,* two possibilities were suggested as alternatives to large operational tests: operational testing carried out in stages and small-scale pilot tests. In this section, we discuss how these ideas might be implemented by ATEC.

We have classified the two basic objectives of operational testing as learning what a system is (and is not) capable of doing in a realistic operational setting, and confirming that a new system's performance is at a certain level or outperforms a baseline system. Addressing these two types of objectives in stages seems natural, with the objective at the first stage being to learn about system performance and the objective at the second stage to confirm a level of system performance.

An operational test could be phased to take advantage of this approach: the first phase might be to examine IBCT/Stryker under different conditions, to assess when this system works best and why. The second phase would be used to compare IBCT/Stryker with a baseline; it would serve as the confirmation experiment used to support the decision to proceed to full-rate production. In the second phase, IBCT/Stryker would be compared with the baseline only in the best and worst scenarios. This broad testing strategy is used by many companies in the pharmaceutical industry and is more fully described in Box, Hunter, and Hunter (1978).

Some of the challenges now faced by ATEC result from an attempt to simultaneously address the two objectives of learning and confirming. Clearly, they will often require very different designs. Although there are pragmatic reasons why a multistage test may not be feasible (e.g., difficulty reserving test facilities and scheduling soldiers to carry out the test missions), if these reasons can be addressed the multistage approach has substantial advantages. For example, since TRADOC already conducts some of the learning phase, their efforts could be better integrated with those of ATEC. Also, a multistage process would have implications for how development testing is carried out, especially with respect to the need to have developmental testing make use of as much operational realism as possible, and to have the specific circumstances of developmental test events documented and archived for use by ATEC. An important advantage of this overall approach is that the final operational test may turn out to be smaller than is currently the case.

When specific performance or capability questions come up in the early part of operational testing, small-scale pilot tests, focused on the analysis of these questions, should be seriously considered. For example, the value of situation awareness is not directly addressed by the current operational test for IBCT/Stryker (unless the six additional missions identified in the test plan are used for this purpose). It would be very informative to use Stryker with situation awareness degraded or "turned off" to determine the value that it provides in particular missions (see Chapter 3).

COMMENTS ON THE CURRENT DESIGN IN THE CONTEXT OF CURRENT ATEC CONSTRAINTS

Using the arguments developed above and referring to the current design of the operational test as described in Chapter 2 (and illustrated in Table 2-1), the discussion that follows takes into account the following constraints of the current test design:

1. Essentially no information about the performance of IBCT/Stryker or the baseline has been used to impact the allocation of test samples in the test design.
2. The allocation of test samples to scenarios is constrained to reflect the allocation of use detailed in the OMS/MP.
3. Operational tests are designed to test the system for typical stresses that will be encountered in the field.
4. Operational tests are single test events.

Balanced Design

The primary advantage of the current operational test design is that it is balanced. This means that the test design covers the design space in a systematic and relatively uniform manner (specifically, three security operations in a stable environment, SOSE, appear for every two perimeter defense missions). It is a robust design, in that the test will provide direct, substantial information from all parts of the design space, reducing the need to extrapolate. Even with moderate amounts of missing data, resulting from an inability to carry out a few missions, some information will still be available from all design regions.

Furthermore, if there are no missing data, the balance will permit straightforward analysis and presentation of the results. More specifically, estimation of the effect of any individual factor can be accomplished by collapsing the test results over the remaining factors. And, since estimates of the design effects are uncorrelated in this situation, inference for one effect does not depend on others. However, many of these potential advantages of balance can be lost if there are missing data. If error variances turn out to be heterogeneous, the balanced design will be inefficient compared with a design that would have a priori accommodated the heterogeneity.

The primary disadvantage of the current design is that there is a very strong chance that observed differences will be confounded by important sources of uncontrolled variation. The panel discussed one potential source of confounding in its October 2002 letter report (see Appendix A), which recommends that the difference in starting time between the IBCT/Stryker test missions and the baseline test missions be sufficiently shortened to reduce any differences that seasonal changes (e.g., in foliage and temperature) might cause. Other potential sources of confounding include: (1) player differences due to learning, fatigue, training, and overall competence; (2) weather differences (e.g., amount of precipitation); and (3) differences between IBCT/Stryker and the baseline with respect to the number of daylight and nighttime missions.

In addition, the current design is not *fully* orthogonal (or balanced), which is evident when the current design is collapsed over scenarios. For example, for company B in the SOSE mission type, the urban missions have higher intensity than the rural missions. (After this was brought to the attention of ATEC they were able to develop a fully balanced design, but they were too far advanced in the design phase to implement this change). While the difference between the two designs appears to be small in this particular case, we are nevertheless disappointed that the best possible techniques are not being used in such an important program. This is an indication of the need for access (in this case *earlier* access) to better statistical expertise in the Army test community, discussed in Chapter 6 (as well as in National Research Council, 1998a).

During the operational test, the time of day at which each mission begins is recorded, providing some possibility of checking for player learning and player fatigue. One alternative to address confounding due to player learning is to use four separate groups of players, one for each of the two

OPFORs, one for the IBCT/Stryker, and one for baseline system. Intergroup variability appears likely to be a lesser problem than player learning. Alternating teams from test replication to test replication between the two systems under test would be a reasonable way to address differences in learning, training, fatigue, and competence. However, we understand that either idea might be very difficult to implement at this date.[2]

The confounding factor of extreme weather differences between Stryker and the baseline system can be partially addressed by postponing missions during heavy weather (although this would prevent gaining an understanding of how the system operates in those circumstances). Finally, the lack of control for daylight and nighttime missions remains a concern. It is not clear why this variable could not have been included as a design variable.

Aside from the troublesome confounding issue (and the power calculations commented on below), the current design is competent from a statistical perspective. However, measures to address the various sources of confounding need to be seriously considered before proceeding.

Comments on the Power Calculations[3]

ATEC designed the IBCT/Stryker IOT to support comparisons of the subject-matter expert (SME) ratings between IBCT/Stryker and the baseline for particular types of missions—for example, high-intensity urban missions and medium-intensity rural SOSE missions. In addition, ATEC designed the operational test for IBCT/Stryker to support comparisons relative to attrition at the company level. ATEC provided analyses to justify the assertion that the current test design has sufficient power to support some of these comparisons. We describe these analyses and provide brief comments below.

SME ratings are reported on a subjective scale that ranges from 1 to 8. SMEs will be assigned randomly, with one SME assigned per company, and two SMEs assigned to each platoon mission. SMEs will be used to evaluate mission completion, protection of the force, and avoidance of col-

---

[2]It is even difficult to specify exactly what one would mean by "equal training," since the amount of training needed for the IBCT to operate Stryker is different from that for a Light Infantry Brigade.

[3]The source for this discussion is U.S. Department of Defense (2002b).

lateral damage, which results in 10 to 16 comparisons per test. Assuming that the size of an individual significance test was set equal to 0.01, and that there are 10 different comparisons that are likely to be made, from a Bonferroni-type argument the overall size of the significance tests would be at most 0.1. In our view this control of individual errors is not crucial, and ATEC should instead examine two or three key measures and carry out the relevant comparisons with the knowledge that the overall type I error may be somewhat higher than the stated significance level.

Using previous experience, ATEC determined that it was important to have sufficient power to detect an average SME rating difference of 1.0 for high-intensity missions, 0.75 for medium-intensity missions, 0.50 for low-intensity missions, and 0.75 difference overall. (We have taken these critical rating differences as given, because we do not know how these values were justified; we have questioned above the allocation of test resources to low-intensity missions.) ATEC carried out simulations of SME differences to assess the power of the current operational test design for IBCT/Stryker. While this is an excellent idea in general, we have some concerns as to how these particular simulations were carried out.

First, due to the finite range of the ratings difference distribution, ATEC expressed concern that the nonnormality of *average* SME ratings differences (in particular, the short tail of its distribution) may affect the coverage properties of any confidence intervals that were produced in the subsequent analysis. We are convinced that even with relatively small sample sizes, the means of SME rating differences will be well represented by a normal distribution as a result of the structure of the original distribution and the central limit theorem, and that taking the differences counters skewness effects. Therefore the nonnormality of SME ratings differences should not be a major concern.

ATEC reports that they collected "historical task-rating differences" and determined that the standard deviation of these differences was 1.98, which includes contributions from both random variation and variation in performance between systems. Then ATEC modeled SME ratings scores for both IBCT/Stryker and the baseline using linear functions of the controlled variables from the test design. These linear functions were chosen to produce SME scores in the range between 1 and 8. ATEC then added to these linear functions a simulated random error variation of +1, 0, and −1, each with probability 1/3. The resulting SME scores were then truncated to make them integral (and to lie between 1 and 8). The residual standard error of *differences* of these scores was then estimated, using simulation, to be 1.2.[4] In addition, SME ratings differences (which include random variation as well as modeled performance differences) were simulated, with a resulting observed standard deviation of 2.04. Since this value was close enough to the historical value of 1.98, it supported their view that the amount of random variation added was similar to what would be observed for SMEs in the field.

The residual standard error of the mean is defined to be the residual standard error divided by the square root of the sample size. So, when the test sample size that can be used for a comparison is 36 (essentially the entire operational test minus the 6 additional missions), the residual standard error of the mean will be 0.20; twice that is 0.40. ATEC's analysis argues that since 0.75 is larger than 0.40, the operational test will have sufficient statistical power to find a difference of 0.75 in SME ratings. The same argument was used to show that interaction effects that are estimated using test sample sizes of 18 or 12 would also have sufficient statistical power, but interaction effects that were estimated using test sample sizes of 6 or 4 would not have sufficient statistical power to identify SME ratings differences of 0.75. Furthermore, if the residual standard error of ratings differences were as high as 1.4, a sample size of 12 would no longer provide sufficient power to identify a ratings difference of 0.75.

Our primary concern with this analysis is that the random variation of SME scores has not been estimated directly. It is not clear why SME rating differences would behave similarly to the various historic measures (see Chapter 3). It would have been preferable to run a small pilot study to provide preliminary estimates of these measures and their variance. If that is too expensive, ATEC should identify those values for which residual standard errors provide sufficient power at a number of test sample sizes, as a means of assessing the sensitivity of their analysis to the estimation of these standard errors. (ATEC's point about the result when the residual standard deviation is raised to 1.4 is a good start to this analysis.)

ATEC has suggested increasing statistical power by combining the ratings for a company mission, or by combining ratings for company and platoon missions. We are generally opposed to this idea if it implies that the *uncombined* ratings will not also be reported.

---

[4]For this easy example, simulation was not needed, but simulation might be required in more complicated situations.

During various missions in the IBCT/Stryker operational test, the number of OPFOR players ranges from 90 to 220, and the number of noncombatant or blue forces is constant at 120. Across 36 missions, there are 10,140 potential casualties. For a subset of these (e.g., blue force players), the potential casualties range from 500 to 4,320. ATEC offered analysis asserting that with casualty rates of 13 percent for the baseline and 10 percent for IBCT/Stryker, it will be possible to reject the null hypothesis of equal casualty rates for the two systems under test with statistical power greater than 75 percent. It is not clear what distribution ATEC has assumed for casualty counts, but likely candidates are binomial and Poisson models.

That analysis may be flawed in that it makes use of an assumption that is unlikely to hold: that individual casualties are independent of one another. Clearly, battles that go poorly initially are likely to result in more casualties, due to a common conditioning event that makes individual casualty events dependent. As a result, these statistical power calculations are unlikely to be reliable. Furthermore, not only are casualties not independent, but even if they were, they should not be rolled up across mission types. For example, with respect to assessment of the value of IBCT/Stryker, one casualty in a perimeter defense mission does not equal one casualty in a raid.

The unit of analysis appears to be a complicated issue in this test. For example, the unit of analysis is assumed by ATEC to be a mission or a task for SMEs, but it is assumed to be an individual casualty for the casualty rate measures. Both positions are somewhat extreme. The mission may in some cases be too large to use as the unit of analysis. Individual skirmishes and other events occurring within a mission could be assumed to be relatively independent and objectively assessed or measured, either by SMEs or by instrumentation. In taking this intermediate approach, the operational test could be shown to have much greater power to identify various differences than the SME analysis discussed above indicates.

Finally, we note that although the current operational test tests only company-level operations, brigade-level testing could be accomplished by using one real brigade-level commander supported by (a) two real battalion commanders, each supported by one real company and two simulated companies and (b) one simulated battalion commander supported by three simulated companies.

SUMMARY

It is inefficient to discover major design flaws during an operational test that could have been discovered earlier in developmental test. Operational test should instead focus its limited sample size on providing operationally relevant information sufficient to support the decision of whether to proceed to full-rate production, and sufficient to refine the system design to address operationally relevant deficiencies. The current design for the IBCT/Stryker operational test is driven by the overall goal of testing the average difference, but it is not as effective at providing information for different scenarios of interest. The primary disadvantage of the current design, in the context of current ATEC constraints, is that there is a distinct possibility that observed differences will be confounded by important sources of uncontrolled variation (e.g., factors associated with seasonal differences).

In the panel's view, it would be worthwhile for ATEC to consider a number of changes in the IBCT/Stryker test design:

1. ATEC should consider, for future test designs, relaxing various rules of test design that it adheres to, by (a) not allocating sample size to scenarios to reflect the OMS/MP, but instead using principles from optimal experimental design theory to allocate sample size to scenarios, (b) testing under somewhat more extreme conditions than typically will be faced in the field, (c) using information from developmental testing to improve test design, and (d) separating the operational test into at least two stages, learning and confirmatory.

2. ATEC should consider applying to future operational testing in general a two-phase test design that involves, first, learning phase studies that examine the test object under different conditions, thereby helping testers design further tests to elucidate areas of greatest uncertainty and importance, and, second, a phase involving confirmatory tests to address hypotheses concerning performance vis-à-vis a baseline system or in comparison with requirements. ATEC should consider taking advantage of this approach for the IBCT/Stryker IOT. That is, examining in the first phase IBCT/Stryker under different conditions, to assess when this system works best, and why, and conducting a second phase to compare IBCT/Stryker to a baseline, using this confirmation experiment to support the decision to proceed to full-rate production. An important feature of the learning phase

is to test with factors at high stress levels in order to develop a complete understanding of the system's capabilities and limitations.

3. When specific performance or capability problems come up in the early part of operational testing, small-scale pilot tests, focused on the analysis of these problems, should be seriously considered. For example, ATEC should consider test conditions that involve using Stryker with situation awareness degraded or turned off to determine the value that it provides in particular missions.

4. ATEC should eliminate from the IBCT/Stryker IOT one significant potential source of confounding, seasonal variation, in accordance with the recommendation provided earlier in the October 2002 letter report from the panel to ATEC (see Appendix A). In addition, ATEC should also seriously consider ways to reduce or eliminate possible confounding from player learning, and day/night imbalance. One possible way of addressing the concern about player learning is to use four separate groups of players for the two OPFORs, the IBCT/Stryker, and the baseline system. Also, alternating teams from test replication to test replication between the two systems under test would be a reasonable way to address differences in learning, training, fatigue, and competence.

5. ATEC should reconsider for the IBCT/Stryker their assumption concerning the distribution of SME scores and should estimate the residual standard errors directly, for example, by running a small pilot study to provide preliminary estimates; or, if that is too expensive, by identifying those SME score differences for which residual standard errors provide sufficient power at a number of test sample sizes, as a means of assessing the sensitivity of their analysis to the estimation of these standard errors.

6. ATEC should reexamine their statistical power calculations for the IBCT/Stryker IOT, taking into account the fact that individual casualties may not be independent of one another.

7. ATEC should reconsider the current units of analysis for the IBCT/Stryker testing—a mission or a task for SME ratings, but an individual casualty for the casualty rate measures. For example, individual skirmishes and other events that occur within a mission should be objectively assessed or measured, either by SMEs or by instrumentation.

8. Given either a learning or a confirmatory objective, *ignoring various tactical considerations*, a requisite for operational testing is that it should not commence until the system design is mature.

Finally, to address the limitation that the current IBCT/Stryker IOT tests only company-level operations, ATEC might consider brigade-level testing, for example, by using one real brigade-level commander supported by (a) two real battalion commanders, each supported by one real company and two simulated companies, and (b) one simulated battalion commander supported by three simulated companies.

5

Data Analysis

The panel has noted (see the October 2002 letter report in Appendix A) the importance of determining, prior to the collection of data, the types of results expected and the data analyses that will be carried out. This is necessary to ensure that the designed data collection effort will provide enough information of the right types to allow for a fruitful evaluation. Failure to think about the data analysis prior to data collection may result in omitted explanatory or response variables or inadequate sample size to provide statistical support for important decisions.

Also, if the questions of interest are not identified in advance but instead are determined by looking at the data, then it is not possible to formally address the questions, using statistical arguments, until an independent confirmatory study is carried out.

An important characteristic of the IBCT/Stryker IOT, probably in common with other defense system evaluations, is that there are a large number of measures collected during the evaluation. This includes measures of a variety of types (e.g., counts of events, proportions, binary outcomes) related to a variety of subjects (e.g., mission performance, casualties, reliability). In addition, there are a large number of questions of interest. For the IBCT/Stryker IOT, these include: Does the Stryker-equipped force outperform a baseline force? In which situations does the Stryker-equipped force have the greatest advantage? *Why* does the Stryker-equipped force outperform the baseline force? It is important to avoid "rolling up" the many measures into a small number of summary measures focused only on certain preidentified critical issues. Instead, appropriate measures should be used to address each of the many possible questions. It will sometimes, but certainly not always, be useful to combine measures into an overall summary measure.

The design discussion in Chapter 4 introduced the important distinction between the learning phase of a study and the confirmatory phase of a study. There we recommend that the study proceed in steps or stages rather than as a single large evaluation. This section focuses on the analysis of the data collected. The comments here are relevant whether a single evaluation test is done (as proposed by ATEC) or a series of studies are carried out (as proposed by the panel).

Another dichotomy that is relevant when analyzing data is that between the use of formal statistical methods (like significance tests) and the use of exploratory methods (often graphical). Formal statistical tests and procedures often play a large role in confirmatory studies (or in the confirmatory phase described in Chapter 4). Less formal methods, known as exploratory analysis, are useful for probing the data to detect interesting or unanticipated data values or patterns. Exploratory analysis is used here in the broad sense, to include but not to be limited by the methods described in Tukey (1977). Exploratory methods often make extensive use of graphs to search for patterns in the data. Exploratory analysis of data is always a good thing, whether the data are collected as part of a confirmatory study to compare two forces or as part of a learning phase study to ascertain the limits of performance for a system.

The remainder of this chapter reviews the general principles behind the formal statistical procedures used in confirmatory studies and those methods used in exploratory statistical analyses and then presents some specific recommendations for data analysis for the IBCT/Stryker IOT.

PRINCIPLES OF DATA ANALYSIS

Formal Statistical Methods in Confirmatory Analyses

A key component of any defense system evaluation is the formal comparison of the new system with an appropriately chosen baseline. It is usually assumed that the new system will outperform the baseline; hence this portion of the analysis can be thought of as confirmatory. Depending on the number of factors incorporated in the design, the statistical assessment could be a two-sample comparison (if there are no other controlled experimental or measured covariate factors) or a regression analysis (if there are other factors). In either case, statistical significance tests or confidence intervals are often used to determine if the observed improvement provided by the new system is too large to have occurred by chance.

Statistical significance tests are commonly used in most scientific fields as an objective method for assessing the evidence provided by a study. The National Research Council (NRC) report *Statistics, Testing, and Defense Acquisition* reviews the role and limitations of significance testing in defense testing (National Research Council, 1998a). It is worthwhile to review some of the issues raised in that report. One of the limitations of significance testing is that it is focused on binary decisions: the null hypothesis (which usually states that there is no difference between the experimental and baseline systems) is rejected or not. If it is rejected, then the main goal of the evaluation is achieved, and the data analysis may move to an exploratory phase to better understand when and why the new system is better. A difficulty with the binary decision is that it obscures information about the size of the improvement afforded by the new system, and it does not recognize the difference between statistical significance and practical significance. The outcome of a significance test is determined both by the amount of improvement observed and by the sample size. Failure to find a statistically significant difference may be because the observed improvement is less than anticipated or because the sample size was not sufficient. Confidence intervals that combine an estimate of the improvement provided by the new system with an estimate of the uncertainty or variability associated with the estimate generally provide more information. Confidence intervals provide information about whether the hypothesis of "no difference" is plausible given the data (as do significance tests) but also inform about the likely size of the improvement provided by the system and its practical significance. Thus confidence intervals should be used with or in place of significance tests.

Other difficulties in using and interpreting the results of significance tests are related to the fact that the two hypotheses are not treated equally. Most significance test calculations are computed under the assumption that the null hypothesis is correct. Tests are typically constructed so that a rejection of the null hypothesis confirms the alternative that we believe (or hope) to be true. The alternative hypothesis is used to suggest the nature of the test and to define the region of values for which the null hypothesis is rejected. Occasionally the alternative hypothesis also figures in statistical power calculations to determine the minimum sample size required in order to be able to detect differences of practical significance. Carrying out tests in this way requires trading off the chances of making two possible errors: rejecting the null hypothesis when it is true and failing to reject the null hypothesis when it is false. Often in practice, little time is spent determining the relative cost of these two types of errors, and as a consequence only the first is taken into account and reported.

The large number of outcomes being assessed can further complicate carrying out significance tests. Traditional significance tests often are designed with a 5 or 10 percent error rate, so that significant differences are declared to be in error only infrequently. However, this also means that if formal comparisons are made for each of 20 or more outcome measures, then the probability of an error in one or more of the decisions can become quite high. Multiple comparison procedures allow for control of the experiment-wide error rate by reducing the acceptable error rate for each individual comparison. Because this makes the individual tests more conservative, it is important to determine whether formal significance tests are required for the many outcome measures. If we think of the analysis as comprising a confirmatory and exploratory phase, then it should be possible to restrict significance testing to a small number of outcomes in the confirmatory phase. The exploratory phase can focus on investigating the scenarios for which improvement seems greatest using confidence intervals and graphical techniques. In fact, we may know in advance that there are some scenarios for which the IBCT/Stryker and baseline performance will not differ, for example, in low-intensity military operations; it does not make sense to carry out significance tests when we expect that the null hypothesis is true or nearly true.

It is also clearly important to identify the proper unit of analysis in carrying out statistical analyses. Often data are collected at several different levels in a study. For example, one might collect data about individual soldiers (especially casualty status), platoons, companies, etc. For many outcome measures, the data about individual soldiers will not be independent, because they share the same assignment. This has important implications for data analysis in that most statistical methods require independent observations. This point is discussed in Chapter 4 in the context of study design and is revisited below in discussing data analysis specifics for the IBCT/Stryker IOT.

Exploratory Analyses

Conclusions obtained from the IOT should not stop with the confirmation that the new system performs better than the baseline. Operational tests also provide an opportunity to learn about the operating characteristics of new systems/forces. Exploratory analyses facilitate learning by making use of graphical techniques to examine the large number of variables and scenarios. For the IBCT/Stryker IOT, it is of interest to determine the factors (mission intensity, environment, mission type, and force) that impact IBCT/Stryker and the effects of these factors. Given the large number of factors and the many outcome measures, the importance of the exploratory phase of the data analysis should not be underestimated.

In fact, it is not even correct to assume (as has been done in this chapter) that formal confirmatory tests will be done prior to exploratory data analysis. Examination of data, especially using graphs, can allow investigators to determine whether the assumptions required for formal statistical procedures are satisfied and identify incorrect or suspect observations. This ensures that appropriate methodology is used in the important confirmatory analyses. The remainder of this section assumes that this important part of exploratory analysis has been carried out prior to the use of formal statistical tests and procedures. The focus here is on another crucial use of exploratory methods, namely, to identify data patterns that may suggest previously unseen advantages or disadvantages for one force or the other.

Tukey (1977) and Chambers et al. (1983) describe an extensive collection of tools and examples for using graphical methods in exploratory data analysis. These methods provide a mechanism for looking at the data to identify interesting results and patterns that provide insight about the system under study. Graphs displaying a single outcome measure against a variety of factors can identify subsets of the design space (i.e., combinations of factors) for which the improvement provided by a new system is noticeably high or low. Such graphs can also identify data collection or recording errors and unexpected aspects of system performance.

Another type of graphical display presents several measures in a single graph (for example, parallel box plots for the different measures or the same measures for different groups). Such graphs can identify sets of outcome measures that show the same pattern of responses to the factors, and so can help confirm either that these measures are all correlated with mission success as expected, or may identify new combinations of measures worthy of consideration. When an exploratory analysis of many independent measures shows results consistent with a priori expectations but not statistically significant, these results might in combination reinforce one another if they could all be attributed to the same underlying cause.

It should be pointed out that exploratory analysis can include formal multivariate statistical methods, such as principal components analysis, to determine which measures appear to correlate highly across mission scenarios (see, for example, Johnson and Wichern, 1992). One might identify combinations of measures that appear to correlate well with the ratings of SMEs, in this way providing a form of objective confirmation of the implicit combination of information done by the experts.

Reliability and Maintainability

These general comments above regarding confirmatory and exploratory analysis apply to all types of outcome measures, including those associated with reliability and maintainability, although the actual statistical techniques used may vary. For example, the use of exponential or Weibull data models is common in reliability work, while normal data models are often dominant in other fields. Meeker and Escobar (1998) provide an excellent discussion of statistical methods for reliability.

A key aspect of complex systems like Stryker that impacts reliability, availability, and maintainability data analysis is the large number of failure modes that affect reliability and availability (discussed also in Chapter 3). These failure modes can be expected to have different behavior. Failure modes due to wear would have increasing hazard over time, whereas other modes would have decreasing hazard over time (as defects are fixed). Rather than using statistical models to directly model system-wide failures, each of the major failure modes should be modeled. Inferences about system-wide reliability would then be obtained by combining information from the different modes.

Thinking about exploratory analysis for reliability and maintainability data raises important issues about data collection. Data regarding the reliability of a vehicle or system should be collected from the start of operations and tracked through the lifetime of the vehicle, including training uses of the vehicle, operational tests, and ultimately operational use. It is a challenge to collect data in this way and maintain it in a common database, but the ability to do so has important ramifications for reliability modeling. It is also important to keep maintenance records as well, so that the times between maintenance and failures are available.

Modeling and Simulation

Evaluation plans often rely on modeling and simulation to address several aspects of the system being evaluated. Data from the operational test may be needed to run the simulation models that address some issues, but certainly not all; for example, no new data are needed for studying transportability of the system. Information from an operational test may also identify an issue that was not anticipated in pretest simulation work, and this could then be used to refine or improve the simulation models.

In addition, modeling and simulation can be used to better understand operational test results and to extrapolate to larger units. This is done by using data from the operational test to recreate and/or visualize test events. The recreated events may then be further probed via simulation. In addition, data (e.g., on the distributions of events) can be used to run through simulation programs and assess factors likely to be important at the brigade level. Care should be taken to assess the uncertainty effect of the limited sample size results from the IOT on the scaled-up simulations.

ANALYSIS OF DATA FROM THE IBCT/STRYKER IOT

This section addresses more specifically the analysis of data to be collected from the IBCT/Stryker IOT. Comments here are based primarily on information provided to the panel in various documents (see Chapter 1) and briefings by ATEC that describe the test and evaluation plans for the IBCT/Stryker.

Confirmatory Analysis

ATEC has provided us with detailed plans describing the intended analysis of the SME scores of mission outcomes and mission casualty rates. These plans are discussed here.

The discussion of general principles in the preceding section comments on the importance of defining the appropriate unit for data analysis. The ATEC-designed evaluation consists basically of 36 missions for the Stryker-equipped force and 36 missions for the baseline force (and the 6 additional missions in the ATEC design reserved for special studies). These missions are defined by a mix of factors, including mission type (raid, perimeter defense, area presence), mission intensity (high, medium, low), location (rural, urban), and company pair (B, C). The planned analysis of SME mission scores uses the mission as the basic unit. This seems reasonable, although it may be possible to carry out some data analysis using company-level or platoon-level data or using events within missions (as described in Chapter 4). The planned analysis of casualty rates appears to work with the individual soldier as the unit of analysis. In the panel's view this is incorrect because there is sure to be dependence among the outcomes for different soldiers. Therefore, a single casualty rate should be computed for each mission (or for other units that might be deemed to yield independent information) and these should be analyzed in the manner currently planned for the SME scores.

Several important data issues should be considered by ATEC analysts. These are primarily related to the SME scores. Confirmatory analyses are often based on the assumptions that there is a continuous or at least ordered categorical measurement scale (although they are often done with Poisson or binomial data) and that the measurements on that scale are subject to measurement error that has constant variance (independent of the measured value). The SME scores provide an ordinal scale such that a mission success score of 8 is better than a score of 7, which is better than a score of 6. It is not clear that the scale can be considered an interval scale in which the difference between an 8 and 7 and between a 7 and 6 are the same. In fact, anecdotal evidence was presented to the panel suggesting that scores 5 through 8 are viewed as successes, and scores 1 through 4 are viewed as failures, which would imply a large gap between 4 and 5. One might also expect differences in the level of variation observed at different points along the scale, for two reasons. First, data values near either end of a scale (e.g., 1 or 8 in the present case) tend to have less measurement variation than those in the middle of the scale. One way to argue this is to note that all observers are likely to agree on judgments of missions with scores of 7 or 8, while there may be more variation on judgments about missions in the middle of the scoring scale (one expert's 3 might be another's 5). Second, the missions are of differing length and complexity. It is quite likely that the scores of longer missions may have more variability than those of shorter missions. Casualty rates, as proportions, are also likely to exhibit nonconstant variance. There is less variation in a low casualty rate (or an extremely high one) and more variation for a casualty rate away from the extremes. Transformations of SME scores or casualty rates should be considered if nonconstant variance is determined to be a problem.

The intended ATEC analysis focuses on the difference between IBCT/Stryker force outcomes and baseline force outcomes for the 36 missions.

By working with differences, the main effects of the various factors are eliminated, providing for more precise measurement of system effectiveness. Note that variation due to interactions, that is to say variation in the benefits provided by IBCT/Stryker over different scenarios, must be addressed through a statistical model. The appropriate analysis, which appears to be part of ATEC plans, is a linear model that relates the difference scores (that is, the difference between the IBCT and baseline performance measures on the same mission) to the effects of the various factors. The estimated residual variance from such a model provides the best estimate of the amount of variation in outcome that would be expected if missions were repeated under the same conditions. This is not the same as simply computing the variance of the 36 differences, as that variance would be inflated by the degree to which the IBCT/Stryker advantage varies across scenarios. The model would be likely to be of the form

D_i = difference score for mission i = overall mean + mission type effect + mission intensity effect + location effect + company effect + other desired interactions + error

The estimated overall mean is the average improvement afforded by IBCT/Stryker relative to the baseline. The null hypothesis of no difference (overall mean = 0) would be tested using traditional methods. Additional parameters measure the degree to which IBCT/Stryker improvement varies by mission type, mission intensity, location, company, etc. These additional parameters can be tested for significance or, as suggested above, estimates for the various factor effects can be reported along with estimates of their precision to aid in the judgment of practically significant results. This same basic model can be applied to other continuous measures, including casualty rate, subject to earlier concerns about homogeneity of variance.

This discussion ignores the six additional missions for each force. These can also be included and would provide additional degrees of freedom and improved error variance estimates.

Exploratory Analysis

It is anticipated that IBCT/Stryker will outperform the baseline. Assuming that result is obtained, the focus will shift to determining under which scenarios Stryker helps most and why. This is likely to be determined by careful analysis of the many measures and scenarios. In particular, it seems valuable to examine the IBCT unit scores, baseline unit scores, and differences graphically to identify any unusual values or scenarios. Such graphical displays will complement the results of the confirmatory analyses described above.

In addition, the exploratory analysis provides an opportunity to consider the wide range of measures available. Thus, in addition to SME scores of mission success, other measures (as described in Chapter 3) could be used. By looking at graphs showing the relationship of mission outcome and factors like intensity simultaneously for multiple outcomes, it should be possible to learn more about IBCT/Stryker's strengths and vulnerabilities. However, the real significance of any such insights would need to be confirmed by additional testing.

Reliability and Maintainability

Reliability and maintainability analyses are likely to be focused on assessing the degree to which Stryker meets the design specifications. Traditional reliability methods will be useful in this regard. The general principles discussed earlier concerning separate modeling for different failure modes is important. It is also important to explore the reliability data across vehicle types to identify groups of vehicles that may share common reliability profiles or, conversely, those with unique reliability problems.

Modeling and Simulation

ATEC has provided little detail about how the IBCT/Stryker IOT data might be used in post-IOT simulations, so we do not discuss this issue. This leaves open the question of whether and how operational test data can be extrapolated to yield information about larger scale operations.

SUMMARY

The IBCT/Stryker IOT is designed to serve two major purposes: (1) confirmation that the Stryker-equipped force will outperform the Light Infantry Brigade baseline, and estimation of the amount by which it will outperform and (2) exploring the performance of the IBCT to learn about the performance capabilities and limitations of Stryker. Statistical significance tests are useful in the confirmatory analysis comparing the Stryker-equipped and baseline forces. In general, however, the issues raised by the 1998 NRC panel suggest that more use should be made of estimates and associated measures of precision (or confidence intervals) in addition to significance tests because the former enable the judging of the practical significance of observed effects. There is a great deal to be learned by exploratory analysis of the IOT data, especially using graphical methods. The data may instruct ATEC about the relative advantage of IBCT/Stryker in different scenarios as well as any unusual events during the operational test.

We call attention to several key issues:

1. The IBCT/Stryker IOT involves the collection of a large number of measures intended to address a wide variety of issues. The measures should be used to address relevant issues without being rolled up into overall summaries until necessary.

2. The statistical methods to be used by ATEC are designed for independent study units. In particular, it is not appropriate to compare casualty rates by simply aggregating indicators for each soldier over a set of missions. Casualty rates should be calculated for each mission (or possibly for discrete events of shorter duration) and these used in subsequent data analyses.

3. The IOT provides little vehicle operating data and thus may not be sufficient to address all of the reliability and maintainability concerns of ATEC. This highlights the need for improved data collection regarding vehicle usage. In particular, data should be maintained for each vehicle over that vehicle's entire life, including training, testing, and ultimately field use; data should also be gathered separately for different failure modes.

4. The panel reaffirms the recommendation of the 1998 NRC panel that more use should be made of estimates and associated measures of precision (or confidence intervals) in addition to significance tests, because the former enable the judging of the practical significance of observed effects.

6

Assessing the IBCT/Stryker Operational Test in a Broad Context

In our work reported here, the Panel on the Operational Test Design and Evaluation of the Interim Armored Vehicle has used the report of the Panel on Statistical Methods for Testing and Evaluating Defense Systems (National Research Council, 1998a, referred to in this chapter as NRC 1998) to guide our thinking about evaluating the IBCT/Stryker Initial Operational Test (IOT). Consistent with our charge, we view our work as a case study of how the principles and practices put forward by the previous panel apply to the operational test design and evaluation of IBCT/Stryker. In this context, we have examined the measures, design, and evaluation strategy of IBCT/Stryker in light of the conclusions and recommendations put forward in NRC 1998 with the goal of deriving more general findings of broad applicability in the defense test and evaluation community.

From a practical point of view, it is clear that several of the ideas put forward in NRC 1998 for improvement of the measures and test design cannot be implemented in the IBCT/Stryker IOT due to various constraints, especially time limitations. However, by viewing the Styker test as an opportunity to gain additional insights into how to do good operational test design and evaluation, our panel hopes to further sharpen and disseminate the ideas contained in NRC 1998. In addition, this perspective will demonstrate that nearly all of the recommendations contained in this report are based on generally accepted principles of test design and evaluation.

Although we note that many of the recommendations contained in NRC 1998 have not been fully acted on by ATEC or by the broader defense test and evaluation community, this is not meant as criticism. The paradigm shift called for in that report could not have been implemented in the short amount of time since it has been available. Instead, our aim is to more clearly communicate the principles and practices contained in NRC 1998 to the broad defense acquisition community, so that the changes suggested will be more widely understood and adopted.

A RECOMMENDED PARADIGM FOR TESTING AND EVALUATION

Operational tests, by necessity, are often large, very complicated, and expensive undertakings. The primary contribution of an operational test to the accumulated evidence about a defense system's operational suitability and effectiveness that exist a priori is that it is the only objective assessment of the interaction between the soldier and the complete system as it will be used in the field. It is well known that a number of failure modes and other considerations that affect a system's performance are best (or even uniquely) exhibited under these conditions. For this reason, Conclusion 2.3 of NRC 1998 states: "operational testing is essential for defense system evaluation."

Operational tests have been put forward as tests that can, in isolation from other sources of information, provide confirmatory statistical "proof" that specific operational requirements have been met. However, a major finding of NRC 1998 is that, given the test size that is typical of the operational tests of large Acquisition Category I (ACAT I) systems and the heterogeneity of the performance of these systems across environments of use, users, tactics, and doctrine, operational tests cannot, generally speaking, satisfy this role.[1] Instead, the test and evaluation process should be viewed as a continuous process of information collection, analysis, and decision making, starting with information collected from field experience of the

---

[1] Conclusion 2.2 of the NRC 1998 report states: "The operational test and evaluation requirement, stated in law, that the Director, Operational Test and Evaluation certify that a system is operationally effective and suitable often cannot be supported solely by the use of standard statistical measures of confidence for complex defense systems with reasonable amounts of testing resources" (p. 33).

baseline and similar systems, and systems with similar or identical components, through contractor testing of the system in question, and then through developmental testing and operational testing (and in some sense continued after fielding forward to field performance).

Perhaps the most widely used statistical method for supporting decisions made from operational test results is significance testing. Significance testing is flawed in this application because of inadequate test sample size to detect differences of practical importance (see NRC, 1998:88-91), and because it focuses attention inappropriately on a pass/fail decision rather than on learning about the system's performance in a variety of settings. Also, significance testing answers the wrong question—not whether the system's performance satisfies its requirements but whether the system's performance is inconsistent with failure to meet its requirements—and significance testing fails to balance the risk of accepting a "bad" system against the risk of rejecting a "good" system. Significance tests are designed to detect statistically significant differences from requirements, but they do not address whether any differences that may be detected are practically significant.

The DoD milestone process must be rethought, in order to replace the fundamental role that significance testing currently plays in the pass/fail decision with a fuller exploration of the consequences of the various possible decisions. Significance tests and confidence intervals[2] provide useful information, but they should be augmented by other numeric and analytic assessments using all information available, especially from other tests and trials. An effective formal decision-making framework could use, for example, significance testing augmented by assessments of the likelihood of various hypotheses about the performance of the system under test (and the baseline system), as well as the costs of making various decisions based on whether the various alternatives are true. Moreover, designs used in operational testing are not usually constructed to inform the actual decisions that operational test is intended to support. For example, if a new system is supposed to outperform a baseline in specific types of environments, the test should provide sufficient test sample in those environments to determine whether the advantages have been realized, if necessary at the

---

[2]Producing confidence intervals for sophisticated estimates often requires resampling methods.

cost of test sample in environments where the system is only supposed to equal the baseline.

Testing the IBCT/Stryker is even more complicated than many ACAT I systems in that it is really a test of a system of systems, not simply a test of what Stryker itself is capable of. It is therefore no surprise that the size of the operational test (i.e., the number of test replications) for IBCT/Stryker will be inadequate to support many significance tests that could be used to base decisions on whether Stryker should be passed to full-rate production. Such decisions therefore need to be supplemented with information from the other sources, mentioned above.

This argument about the role of significance testing is even more important for systems such as the Stryker that are placed into operational testing when the system's performance (much less its physical characteristics) has not matured, since then the test size needs to be larger to achieve reasonable power levels. When a fully mature system is placed into operational testing, the test is more of a confirmatory exercise, a shakedown test, since it is essentially understood that the requirements are very likely to be met, and the test can then focus on achieving a greater understanding of how the system performs in various environments.

Recommendation 3.3 of NRC 1998 argued strongly that information should be used and appropriately combined from all phases of system development and testing, and that this information needs to be properly archived to facilitate retrieval and use. In the case of the IBCT/Stryker IOT, it is clear that this has not occurred, as evidenced by the difficulty ATEC has had in accessing relevant information from contractor testing and, indeed, operational experiences from allies using predecessor systems (e.g., the Canadian LAV-III).

HOW IBCT/STRYKER IOT CONFORMS WITH RECOMMENDATIONS FROM THE NRC 1998 REPORT

Preliminaries to Testing

The new paradigm articulated in NRC 1998 argues that defense systems should not enter into operational testing unless the system design is relatively mature. This maturation should be expedited through previous testing that incorporates various aspects of operational realism in addition to the usual developmental testing. The role, then, for operational testing would be to confirm the results from this earlier testing and to learn more about how to operate the system in different environments and what the system's limitations are. The panel believes that in some important respects Stryker is not likely to be fully ready for operational testing when that is scheduled to begin. This is because:

1. many of the vehicle types have not yet been examined for their suitability, having been driven only a fraction of the required mean miles to failure (1,000 miles);
2. the use of add-on armor has not been adequately tested prior to the operational test;
3. it is still not clear how IBCT/Stryker needs to be used in various types of scenarios, given the incomplete development of its tactics and doctrine; and
4. the GFE systems providing situation awareness have not been sufficiently tested to guarantee that the software has adequate reliability.

The role of operational test as a confirmatory exercise has therefore not been realized for IBCT/Stryker. This does not necessarily mean that the IOT should be postponed, since the decision to go to operational test is based on a number of additional considerations. However, it does mean that the operational test is being run with some additional complications that could reduce its effectiveness.

Besides system maturity, another prerequisite for an operational test is a full understanding of the factors that affect system performance. While ATEC clearly understands the most crucial factors that will contribute to variation in system performance (intensity, urban/rural, day/night, terrain, and mission type), it is not clear whether they have carried out a systematic test planning exercise, including (quoting from NRC, 1998a:64-65): "(1) defining the purpose of the test; . . . (4) using previous information to compare variation within and across environments, and to understand system performance as a function of test factors; . . . and (6) use of small-scale screening or guiding tests for collecting information on test planning."

Also, as mentioned in Chapter 4, it is not yet clear that the test design and the subsequent test analysis have been linked. For example, if performance in a specific environment is key to the evaluation of IBCT/Stryker, more test replications will need to be allocated to that environment. In addition, while the main factors affecting performance have been identified, factors such as season, day versus night, and learning effects were not, at least initially, explicitly controlled for. This issue was raised in the panel's letter report (Appendix A).

Test Design

This section discusses two issues relevant to test design: the basic test design and the allocation of test replications to design cells. First, ATEC has decided to use a balanced design to give it the most flexibility in estimating the variety of main effects of interest. As a result, the effects of terrain, intensity, mission, and scenario on the performance of these systems will be jointly estimated quite well, given the test sample size. However, at this point in system development, ATEC does not appear to know which of these factors matter more and/or less, or where the major uncertainties lie. Thus, it may be that there is only a minority of environments in which IBCT/Stryker offers distinct advantages, in which case those environments could be more thoroughly tested to achieve a better understanding of its advantages in those situations. Specific questions of interest, such as the value of situation awareness in explaining the advantage of IBCT/Stryker, can be addressed by designing and running small side experiments (which might also be addressed prior to a final operational test). This last suggestion is based on Recommendation 3.4 of the NRC 1998 report (p. 49): "All services should explore the adoption of the use of small-scale testing similar to the Army concept of force development test and experimentation."

Modeling and simulation are discussed in NRC 1998 as an important tool in test planning. ATEC should take better advantage of information from modeling and simulation, as well as from developmental testing, that could be very useful for the IBCT/Stryker test planning. This includes information as to when the benefits of the IBCT/Stryker over the baseline are likely to be important but not well established.

Finally, in designing a test, the goals of the test have to be kept in mind. If the goal of an operational test is to learn about system capabilities, then test replications should be focused on those environments in which the most can be learned about how the system's capabilities provide advantages. For example, if IBCT/Stryker is intended primarily as an urban system, more replications should be allocated to urban environments than to rural ones. We understand ATEC's view that its operational test designs must allocate, to the extent possible, replications to environments in accordance with the allocation of expected field use, as presented in the OMS/

MP. In our judgment the OMS/MP need only refer to the operational evaluation, and certainly once estimates of test performance in each environment are derived, they can be reweighted to correspond to summary measures defined by the OMS/MP (which may still be criticized for focusing too much on such summary measures in comparison to more detailed assessments).

Furthermore, there are substantial advantages obtained with respect to designing operational tests by separating the two goals of confirming that various requirements have been met and of learning as much as possible about the capabilities and possible deficiencies of the system before going to full-rate production. That separation allows the designs for these two separate tests to target these two distinct objectives.

Given the recent emphasis in DoD acquisition on spiral development, it is interesting to speculate about how staged testing might be incorporated into this management concept. One possibility is a test strategy in which the learning phase makes use of early prototypes of the subsequent stage of development.

System Suitability

Recommendation 7.1 of NRC 1998 states (p. 105):

> The Department of Defense and the military services should give increased attention to their reliability, availability, and maintainability data collection and analysis procedures because deficiencies continue to be responsible for many of the current field problems and concerns about military readiness.

While criticizing developmental and operational test design as being too focused on evaluation of system effectiveness at the expense of evaluation of system suitability, this recommendation is not meant to suggest that operational tests should be strongly geared toward estimation of system suitability, since these large-scale exercises cannot be expected to run long enough to estimate fatigue life, etc. However, developmental testing can give measurement of system (operational) suitability a greater priority and can be structured to provide its test events with greater operational realism. Use of developmental test events with greater operational realism also should facilitate development of models for combining information, the topic of this panel's next report.

The NRC 1998 report also criticized the test and evaluation community for relying too heavily on the assumption that the interarrival time for initial failures follows an exponential distribution. The requirement for Stryker of 1,000 mean miles between failures makes sense as a relevant measure only if ATEC is relying on the assumption of exponentially distributed times to failure. Given that Stryker, being essentially a mechanical system, will not have exponentially distributed times to failure, due to wearout, the actual distribution of waiting times to failure needs to be estimated and presented to decision makers so that they understand its range of performance. Along the same lines, Stryker will, in all probability, be repaired during the operational test and returned to action. Understanding the variation in suitability between a repaired and a new system should be an important part of the operational test.

Testing of Software-Intensive Systems

The panel has been told that obtaining information about the performance of GFE is not a priority of the IOT: GFE will be assumed to have well-estimated performance parameters, so the test should focus on the non-GFE components of Stryker. One of the components of Stryker's GFE is the software providing Stryker with situation awareness. A primary assumption underlying the argument for the development of Stryker was that the increased vulnerability of IBCT/Stryker (due to its reduced armor) is offset by the benefits gained from the enhanced firepower and defensive positions that Stryker will have due to its greater awareness of the placement of friendly and enemy forces. There is some evidence (FBCB2 test results) that this situation awareness capability is not fully mature at this date. It would therefore not be surprising if newly developed, complex software will suffer reliability or other performance problems that will not be fully resolved prior to the start of operational testing.

NRC 1998 details procedures that need to be more widely adopted for the development and testing of software-intensive systems, including usage-based testing. Further, Recommendation 8.4 of that report urges that software failures in the field should be collected and analyzed. Making use of the information on situation awareness collected during training exercises and in contractor and developmental testing in the operational test design would have helped in the more comprehensive assessment of the performance of IBCT/Stryker. For example, allocating test replications to situations in which previous difficulties in situation awareness had been experienced would have been very informative as to whether the system is effective enough to pass to full-rate production.

Greater Access to Statistical Expertise in Operational Test and Evaluation

Stryker, if fully procured, will be a multibillion dollar system. Clearly, the decision on whether to pass Stryker to full-rate production is extremely important. Therefore, the operational test design and evaluation for Stryker needs to be representative of the best possible current practice. The statistical resources allocated to this task were extremely limited. The enlistment of the National Research Council for high-level review of the test design and evaluation plans is commendable. However, this does not substitute for detailed, hands-on, expert attention by a cadre of personnel trained in statistics with "ownership" of the design and subsequent test and evaluation. ATEC should give a high priority to developing a contractual relationship with leading practitioners in the fields of reliability estimation, experimental design, and methods for combining information to help them in future IOTs. (Chapter 10 of NRC 1998 discusses this issue.)

SUMMARY

The role of operational testing as a confirmatory exercise evaluating a mature system design has not been realized for IBCT/Stryker. This does not necessarily mean that the IOT should be postponed, since the decision to go to operational testing is based on a number of additional considerations. However, it does mean that the operational test is being asked to provide more information than can be expected. The IOT may illuminate potential problems with the IBCT and Stryker, but it may not be able to convincingly demonstrate system effectiveness.

We understand ATEC's view that its operational test designs must allocate, to the extent possible, replications to environments in accordance with the allocation of expected field use, as presented in the OMS/MP. In the panel's judgment, the OMS/MP need only refer to the operational evaluation, and once estimates of test performance in each environment are derived, they can be reweighted to correspond to summary measures defined by the OMS/MP.

We call attention to a number of key points:

1. Operational tests should not be strongly geared toward estimation of system suitability, since they cannot be expected to run long enough to estimate fatigue life, estimate repair and replacement times, identify failure

modes, etc. Therefore, developmental testing should give greater priority to measurement of system (operational) suitability and should be structured to provide its test events with greater operational realism.

2. Since the size of the operational test (i.e., the number of test replications) for IBCT/Stryker will be inadequate to support significance tests leading to a decision on whether Stryker should be passed to full-rate production, ATEC should augment this decision by other numerical and graphical assessments from this IOT and other tests and trials.

3. In general, complex systems should not be forwarded to operational testing, absent strategic considerations, until the system design is relatively mature. Forwarding an immature system to operational test is an expensive way to discover errors that could have been detected in developmental testing, and it reduces the ability of an operational test to carry out its proper function. System maturation should be expedited through previous testing that incorporates various aspects of operational realism in addition to the usual developmental testing.

4. Because it is not yet clear that the test design and the subsequent test analysis have been linked, ATEC should prepare a straw man test evaluation report in advance of test design, as recommended in the panel's October 2002 letter to ATEC (see Appendix A).

5. The goals of the initial operational test need to be more clearly specified. Two important types of goals for operational test are learning about system performance and confirming system performance in comparison to requirements and in comparison to the performance of baseline systems. These two different types of goals argue for different stages of operational test. Furthermore, to improve test designs that address these different types of goals, information from previous stages of system development need to be utilized.

6. To achieve needed detailed, hands-on, expert attention by a cadre of statistically trained personnel with "ownership" of the design and subsequent test and evaluation, the Department of Defense and ATEC in particular should give a high priority to developing a contractual relationship with leading practitioners in the fields of reliability estimation, experimental design, and methods for combining information to help them with future IOTs.

References

Box, G.E.P., Hunter, W.G., and Hunter, J.S.
   1978   *Statistics for Experimenters*. New York: John Wiley & Sons.

Chambers, J.M., Cleveland,W.S., Kleiner, B., and Tukey, P. A.
   1983   *Graphical Methods for Data Analysis*. Belmont, CA: Wadsworth.

Helmbold R.L.
   1992   Casualty Fractions and Casualty Exchange Ratio. Unpublished memorandum to J. Riente, February 12, 1992.

Johnson, R.A. and Wichern, D.W.
   1992   *Applied multivariate statistical analysis*, 3rd edition. Englewood Cliffs, NJ:Prentice Hall.

Meeker, William Q., and Escobar, Luis A.
   1998   *Statistical Methods for Reliability Data*. New York: John Wiley & Sons.

National Research Council
   1998a   *Statistics, Testing, and Defense Acquisition: New Approaches and Methodological Improvements*. Michael L. Cohen, John E. Rolph, and Duane L. Steffey, Eds. Panel on Statistical Methods for Testing and Evaluating Defense Systems, Committee on National Statistics. Washington, DC: National Academy Press.
   1998b   *Modeling Human and Organizational Behavior*. R.W. Pew and A.S. Mavor, Eds. Panel on Modeling Human Behavior and Command Decision Making: Representations for Military Simulations. Washington, DC: National Academy Press.

Thompson, D.
   1992   *The Casualty-FER Curve of the Force Structure Reduction Study: A Comparison to Historical Data*. Vector Research, Incorporated Document No. VRI-OHD WP92-1, March 10, 1992. Ann Arbor, MI: Vector Research, Incorporated.

Tukey, J.W.
   1977   *Exploratory Data Analysis.* New York: Addison-Wesley, .

U.S. Department of Defense
   2000   Operational Requirements Document (ORD) for a Family of Interim Armored Vehicles (IAV), ACAT I, Prepared for the Milestone I Decision, April 6. U.S. Army Training and Doctrine Command, Fort Monroe, Virginia.
   2001   Test And Evaluation Master Plan (TEMP): Stryker Family of Interim Armored Vehicles (IAV). Revision 1, Nov 12. U.S. Army Test and Evaluation Command, Alexandria, Virginia.
   2002a   Interim Armored Vehicle IOTE: Test Design Review with NAS Panel. Unpublished presentation, Nancy Dunn and Bruce Grigg, April 15.
   2002b   Interim Armored Vehicle IOTE: Test Design Review with NAS Panel; Power and Sample Size Considerations. Unpublished presentation, Nancy Dunn and Bruce Grigg, May 6.
   2002c   System Evaluation Plan (SEP) for the Stryker Family of Interim Armored Vehicles (IAV), May 22. U.S. Army Test and Evaluation Command, Alexandria, Virginia.

Veit, Clairice T.
   1996   Judgments in military research: The elusive validity issue. *Phalanx.*

Appendix A

Letter Report of the Panel to the Army Test and Evaluation Command

THE NATIONAL ACADEMIES
Advisers to the Nation on Science, Engineering, and Medicine

Division of Behavioral and Social Sciences and Education
Committee on National Statistics
Panel on Operational Test Design and Evaluation of the
Interim Armored Vehicle (IAV)

500 Fifth Street, NW
Washington, DC 20001
Phone: 202 334 3408
Fax: 202 334 3584
Email:jmcgee@nas.edu

October 10, 2002

Frank John Apicella
Technical Director
Army Evaluation Center
U.S. Army Test and Evaluation Command
4501 Ford Avenue
Alexandria, VA 22302-1458

Dear Mr. Apicella:

As you know, at the request of the Army Test and Evaluation Command (ATEC) the Committee on National Statistics has convened a panel to examine ATEC's plans for the operational test design and evaluation of the Interim Armored Vehicle, now referred to as the Stryker. The panel is currently engaged in its tasks of focusing on three aspects of the operational test design and evaluation of the Stryker: (1) the measures of performance and effectiveness used to compare the Interim Brigade Combat Team (IBCT), equipped with the Stryker, against a baseline force; (2) whether the current operational test design is consistent with state-of-the-art methods in statistical experimental design; and (3) the applicability of models for combining information from testing and field use of related systems and from developmental test results for the Stryker with operational test results for the Stryker. ATEC has asked the panel to comment on ATEC's current plans and to suggest alternatives.

The work performance plan includes the preparation of three reports:

- The first interim report (due in November 2002) will address two topics: (1) the measures of performance and effectiveness used to compare the Stryker-equipped IBCT against the baseline force, and (2) whether the current operational test design is consistent with state-of-the-art methods in statistical experimental design.

- The second interim report (due in March 2003) will address the topic of the applicability of models for combining information from testing and field use of related systems and from developmental test results for the Stryker with operational test results for the Stryker.
- The final report (due in July 2003) will integrate the two interim reports and add any additional findings of the panel.

The reports have been sequenced and timed for delivery to support ATEC's time-critical schedule for developing plans for designing and implementing operational tests and for performing analyses and evaluations of the test results.

Two specific purposes of the initial operational test of the Stryker are to determine whether the Interim Brigade Combat Team (IBCT) equipped with the Stryker performs more effectively than a baseline force (Light Infantry Brigade), and whether the Stryker meets its performance requirements. The results of the initial operational test contribute to the Army's decisions of whether and how to employ the Stryker and the IBCT. The panel's first interim report will address in detail factors relating to the effectiveness and performance of the Stryker-equipped IBCT and of the Stryker; effective experimental designs and procedures for testing these forces and their systems under relevant operational conditions, missions, and scenarios; subjective and objective measures of performance and effectiveness for criteria of suitability, force effectiveness, and survivability; and analytical procedures and methods appropriate to assessing whether and why the Stryker-equipped IBCT compares well (or not well) against the baseline force, and whether and why the Stryker meets (or does not meet) its performance requirements.

In the process of deliberations toward producing the first interim report that will address this broad sweep of issues relevant to operational test design and to measures of performance and effectiveness, the panel has discerned two issues with long lead times to which, in the opinion of the panel, ATEC should begin attending immediately, so that resources can be identified, mustered, and applied in time to address them: early development of a "straw man" (hypothetical draft) Test and Evaluation Report (which will support the development of measures and the test design as well as the subsequent analytical efforts) and the scheduling of test participation by the Stryker-equipped force and the baseline force so as to remove an obvious test confounder of different seasonal conditions.

The general purpose of the initial operational test (IOT) is to provide information to decision makers about the utility of and the remaining challenges to the IBCT and the Stryker system. This information is to be generated through the analysis of IOT results. In order to highlight areas for which data are lacking, the panel strongly recommends that immediate effort be focused on specifying how the test data will be analyzed to address relevant decision issues and questions. Specifically, a straw man Test Evaluation Report (TER) should be prepared as if the IOT had been completed. It should include examples of how the representative data will be analyzed, specific presentation formats (including graphs) with expected results, insights one might develop from the data, draft recommendations, etc. The content of this straw man report should be based on the experience and intuition of the analysts and what they think the results of the IOT might look like. Overall, this could serve to provide a set of hypotheses that would be tested with the actual results. Preparation of this straw man TER will help ATEC assess those issues that cannot be informed by the operational tests as currently planned, will expose areas for which needed data is lacking, and will allow appropriate revision of the current operational test plan.

The current test design calls for the execution of the IBCT/Stryker vs. the opposing force (OPFOR) trials and the baseline vs. the OPFOR trials to be scheduled for different seasons. This design totally confounds time of year with the primary measure of interest: the difference in effectiveness between the baseline force and the IBCT/Stryker force. The panel believes that the factors that are present in seasonal variations—weather, foliage density, light level, temperature, etc.—may have a greater effect on the differences between the measures of the two forces than the abilities of the two forces themselves. We therefore recommend that serious consideration be given to scheduling these events as closely in time as possible. One way to address the potential confounding of seasonal affects, as well as possible effects of learning by blue forces and by the OPFOR, would be to intersperse activities of the baseline force and the IBCT/Stryker force over time.

The panel remains eager to assist ATEC in improving its plans and processes for operational test and evaluation of the IBCT/Stryker. We are grateful for the support and information you and your staff have consistently provided during our efforts to date. It is the panel's hope that delivering to you the recommendations in this brief letter in a timely fashion will encourage ATEC to begin drafting a straw man Test Evaluation Report in time to influence operational test activities and to implement the change in test plan to allow the compared forces to undergo testing in the same season.

Sincerely yours,

Stephen Pollock, *Chair*
Panel on Operational Test Design and
Evaluation of the Interim Armored Vehicle

Appendix B
Force Exchange Ratio, Historical Win Probability, and Winning with Decisive Force

FORCE EXCHANGE RATIO AND HISTORICAL WIN PROBABILITY

For a number of years the Center for Army Analysis (CAA) analyzed historical combat data to determine the relationship between victory and casualties in land combat. The historical data, contained in the CAA Data Base of Battles (1991 version, CDB91) is from a wide range of battle types—durations ranging from hours to weeks, dates ranging from the 1600s to the late 20th century, and forces involving a variety of nationalities. Based on the analysis of these data (and some motivation from Lanchester's square law formulation), it has been demonstrated (see Center for Army Analysis, 1987, and its references) that:

- the probability of an attacker victory[1] is related to a variable called the "defenders advantage" or *ADV*, where *ADV* is a function of force strengths and final survivors; and
- $ADV \cong \ln(FER)$

Since N = threat forces and M = friendly coalition forces in our definition of the force exchange ratio (*FER*), Figure B-1 depicts the historical relationship between the *FER* and the probability of winning, regardless of

---
[1] Probability of a defender victory is the complement.

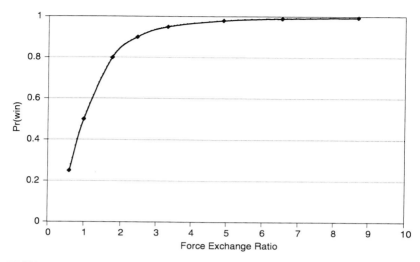

FIGURE B-1 Historical relationship between force exchange ratio and Pr(win). SOURCE: Adapted from Thompson (1992) and Helmbold (1992).

whether the coalition is in defense or attack mode. Additionally, the relation between *FER* and friendly fractional casualties is depicted in Figure B-2 (see CAA, 1992 and VRI, 1992).

FER is not only a useful measure of effectiveness (MOE) to indicate the degree to which force imbalance is reduced, but it is also a useful historical measure of a force's warfighting capability for mission success.

FER AND "DECISIVE FORCE"

Following the demise of the Soviet Union and Operation Desert Storm, the U.S. National Military Strategy (NMS) codified a new military success objective: "Apply decisive force to win swiftly and minimize casualties." The NMS also implied that decisive force will be used to minimize risks associated with regional conflicts. The *FER* is a MOE that is useful in defining and quantifying the level of warfighting capability needed to meet this objective.

Figure B-3 has been derived from a scatterplot of results from a large number of simulated regional conflicts involving joint U.S. forces and coalition partners against a Southwest Asia regional aggressor. The coalition's

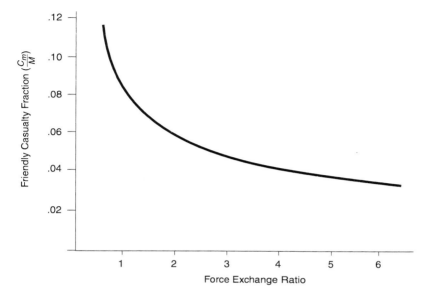

FIGURE B-2 Force exchange ratio/casualty relationship.
SOURCE: Adapted from Thompson (1992) and Helmbold (1992).

objectives are to conduct a defense to prevent the aggressor from capturing critical ports and airfields in Saudi Arabia and to then conduct a counteroffensive to regain lost territory and restore national boundaries.

The *FER*-coalition casualty relationship shown in the figure is based on simulation results, in which the *FER* is the ratio of the percentage of enemy losses to the percentage of coalition losses. Except for the region in which the coalition military objectives were not achieved (*FER* < 1.3) because insufficient forces arrived in the theater, the relationship between *FER* and coalition casualties is similar to that shown in Figure B-2, which is based on historic data. The relationship between *FER* and the probability of win in Figure B-3 is based on the analysis of historic data.

As shown in Figure B-3, a *FER* = 5.0 is defined to be a "decisive" warfighting capability. This level comes close to achieving the criterion of minimizing casualties, since improvements above that level only slightly reduce casualties further. This level of *FER* also minimizes risk in that a force with a *FER* of 2.5 will win approximately 90 out of 100 conflicts (lose 10 percent of the time) but will lose less than 2 percent of the time with a *FER* = 5.0.

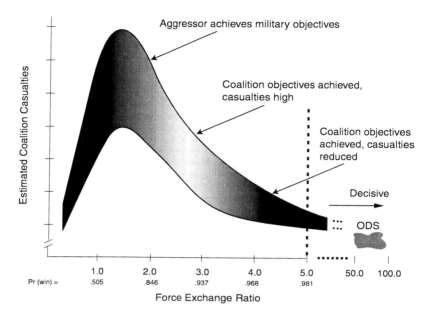

FIGURE B-3 Force exchange ratio and decisive warfighting capability.
SOURCE: Adapted from Thompson (1992) and Helmbold (1992).

Biographical Sketches of Panel Members and Staff

STEPHEN POLLOCK *(Chair)* is Herrick professor of manufacturing in the Department of Industrial and Operations Engineering at the University of Michigan. His research interests are in mathematical modeling, operations research, and Bayesian decision theoretic methods. A former member of the U.S. Army Science Board (1994-1999), he was a member of the National Research Council's (NRC) Committee on Applied and Theoretical Statistics, and he also served on the Panel on Statistical Methods for Testing and Evaluating Defense Systems of the Committee on National Statistics. In addition to his career at the University of Michigan, he spent four years at the U.S. Naval Postgraduate School. He has S.M. and Ph.D. degrees in physics and operations research from the Massachusetts Institute of Technology.

SETH BONDER was until recently chairman and chief executive officer of Vector Research. His area of expertise is military strategy and decision making. He is a member of the National Academy of Engineering, a fellow of the Military Operations Research Society, winner of the Kimball Award from the Operations Research Society of America, recipient of the Award for Patriotic Civilian Service (awarded by the secretary of the Army), and the Vance R. Wanner Memorial Award winner from the Military Operations Research Society. He was a member of the U.S. Defense Science Board summer studies on new operational concepts and on fire support operations. He is a member of the U.S. Army Science Board. He served as a first lieutenant and a captain of the U.S. Air Force. He has a Ph.D. in industrial engineering (operations research) from Ohio State University.

MARION BRYSON is director of research and development at North Tree Fire, International. He was until recently technical director, TEXCOM Experimentation Command, Fort Hood, Texas, and prior to that, director CDEC, Fort Ord, California. He has extensive experience in testing and evaluating defense systems in development. He served as a member of the NRC's Panel on Statistical Methods for Testing and Evaluating Defense Systems of the Committee on National Statistics. He is also a past president and fellow of the Military Operations Research Society, a recipient of the Vance Wanner Memorial Award for leadership in operations research from the Military Operations Research Society, and a recipient of the S.S. Wilks award for research in experimental design. He served as the chief scientist on a task force that developed the requirements for the Apache helicopter, and was a previous associate editor of MORS. He has a Ph.D. in statistics from Iowa State University.

MICHAEL L. COHEN is a senior program officer for the Committee on National Statistics. Previously, he was a mathematical statistician at the Energy Information Administration, an assistant professor in the School of Public Affairs at the University of Maryland, a research associate at the Committee on National Statistics, and a visiting lecturer at the Department of Statistics, Princeton University. He is a fellow of the American Statistical Association. His general area of research is the use of statistics in public policy, with particular interest in census undercount and model validation. He is also interested in robust estimation. He has a B.S. degree in mathematics from the University of Michigan and M.S. and Ph.D. degrees in statistics from Stanford University.

JAMES P. McGEE has been since 1994 a senior project officer at the NRC. In addition to directing this panel, he directs the Panel on Assessing the Scientists and Engineers Statistical Data System of the Committee on National Statistics; projects on health and safety needs of older workers and on susceptibility of older persons to environmental hazards of the Board on Behavioral, Cognitive, and Sensory Sciences; and the panel on soldier systems of the Army Research Laboratory Technical Assessment Board. He has also supported numerous other NRC projects in the areas of human factors psychology, engineering, and education. Prior to joining the NRC, he held technical and management positions as an applied cognitive psychologist at IBM, General Electric, RCA, General Dynamics, and Sikorsky Aircraft corporations. He has a B.A. from Princeton University and a Ph.D. from Fordham University, both in psychology, and for several years instructed postsecondary courses in applied psychology.

WILLIAM Q. MEEKER is distinguished professor of liberal arts and sciences in the Department of Statistics at Iowa State University. His area of expertise is reliability assessment in industrial settings. He is a fellow of the American Statistical Association, an elected member of the International Statistical Institute, a winner of the Frank Wilcoxon prize for the best practical application paper in Technometrics (three times), and a winner of the Youden prize for the best expository paper in Technometrics (also three times). His book (coauthored with Luis A. Escobar) *Statistical Methods for Reliability Data* won the Excellence and Innovation in Engineering Award from the Professional/Scholarly Publishing Division of the Association of American Publishers. He has extensive service as editor and associate editor of *Technometrics* and other distinguished journals. He has a Ph.D. in statistics from Union College.

VIJAYAN NAIR is a professor and currently chair in the Department of Statistics at the University of Michigan. Prior to this, he worked for 12 years as a research scientist at Bell Laboratories. His area of expertise is statistical methods applied to industrial problems, especially in experimental design. He is a fellow of the American Statistical Association, a fellow of the Institute of Mathematical Statistics, an elected member of the International Statistical Institute, and a senior member of the American Society for Quality. He is a winner of the Frank Wilcoxon prize for best practical applications paper in *Technometrics*, and he served as its editor from 1990 to 1992. He was a member of the NRC's Panel on Statistical Methods for Testing and Evaluating Defense Systems of the Committee on National Statistics and also served on the Panel on Information Technology. He has a Ph.D. in statistics from the University of California, Berkeley.

JOHN ROLPH is a professor and chair of the industrial operations and management department in the Marshall School of Business at the University of Southern California. Previously he spent 24 years as a statistician at RAND, 12 of them as head of the statistical research and consulting group. His area of expertise is empirical Bayes estimation and statistics and public policy. He is an elected member of the International Statistical Institute, a fellow of the American Statistical Association, a fellow of the Institute of Mathematical Statistics, and a fellow of the Royal Statistical Society. He has served as a member and now chair of the NRC's Committee on National Statistics, and as member of its Panel on Statistical Methods for Testing and Evaluating Defense Systems. He has also served on three other panels of the NRC. He was for three years editor of *Chance* magazine and in many other editorial capacities. He has a Ph.D. in statistics from the University of California, Berkeley.

FRITZ SCHOLZ is senior statistical analyst in the applied statistics group of The Boeing Company, where he has worked for 23 years. While working for Boeing he has worked on reliability problems in a large variety of industrial settings, including analysis of airplane accident data, analysis of engine failures, modeling of software reliability, nuclear safety, the reliability of rocket motors, nonparametric risk assessment, development of computer models for lightning risk assessment, spares forecasting, and space debris risk assessment. He has contributed to activities related to MIL-HDBK-5 and -17. He was a presenter at the workshop on reliability issues for defense systems held by the NRC's Committee on National Statistics. He is a Boeing technical fellow and a fellow of the American Statistical Association. He has a Ph.D. in statistics from the University of California, Berkeley.

HAL STERN is professor and chair in the Department of Statistics at the University of California, Irvine. His area of expertise is Bayesian methods. He is coauthor of *Bayesian Data Analysis* (with Andrew Gelman, John Carlin, and Donald Rubin). He is a fellow of the American Statistical Association and has served as editor of *Chance* magazine. He has a Ph.D. from Stanford University.

ALYSON WILSON is a technical staff member of the Los Alamos Laboratory Statistical Sciences Group. Prior to that, she was employed by Cowboy Programming Resources, Inc., where she provided analytic support for U.S. Army operational tests of air defense weapon systems. She has written on operational test and evaluation and the use of Bayesian methods. She has a Ph.D. in statistics from Duke University, where her thesis was selected as the winner of the Savage Dissertation Award by the American Statstical Association's section on Bayesian statistics.